Birkhäuser

Trends in Mathematics

Trends in Mathematics is a series devoted to the publication of volumes arising from conferences and lecture series focusing on a particular topic from any area of mathematics. Its aim is to make current developments available to the community as rapidly as possible without compromise to quality and to archive these for reference.

Proposals for volumes can be submitted using the Online Book Project Submission Form at our website www.birkhauser-science.com.

Material submitted for publication must be screened and prepared as follows:

All contributions should undergo a reviewing process similar to that carried out by journals and be checked for correct use of language which, as a rule, is English. Articles without proofs, or which do not contain any significantly new results, should be rejected. High quality survey papers, however, are welcome.

We expect the organizers to deliver manuscripts in a form that is essentially ready for direct reproduction. Any version of TeX is acceptable, but the entire collection of files must be in one particular dialect of TeX and unified according to simple instructions available from Birkhäuser.

Furthermore, in order to guarantee the timely appearance of the proceedings it is essential that the final version of the entire material be submitted no later than one year after the conference.

For further volumes:
http://www.springer.com/series/4961

Concentration Analysis and Applications to PDE

ICTS Workshop, Bangalore, January 2012

Adimurthi
K. Sandeep
Ian Schindler
Cyril Tintarev
Editors

Birkhäuser

Editors
Adimurthi
K. Sandeep
Centre for Applicable Mathematics
Tata Institute of Fundamental Research
Bangalore
Karnataka, India

Ian Schindler
CeReMath
Université Toulouse I
Toulouse, France

Cyril Tintarev
Department of Mathematics
Uppsala University
Uppsala, Sweden

ISBN 978-3-0348-0372-4 ISBN 978-3-0348-0373-1 (eBook)
DOI 10.1007/978-3-0348-0373-1
Springer Basel Heidelberg New York Dordrecht London

Library of Congress Control Number: 2013953867

Mathematics Subject Classification (2010): 35-02, 35-06, 58-02, 58-06, 46-02, 46-06

Printed on acid-free paper

Springer Basel is part of Springer Science+Business Media (www.birkhauser-science.com)

Contents

Concentration Analysis and Applications to PDE
Trends in Mathematics, vii–x

Introduction

Adimurthi, Kunnath Sandeep, Ian Schindler and Cyril Tintarev

The ICTS mini-program *School and workshop on cocompact imbeddings, profile decompositions, and their applications to PDE*[1] took place January 3–12, 2012 in Bangalore, at the Centre for Applicable Mathematics of Tata Institute of Fundamental Research, funded by International Centre for Theoretical Sciences with some additional support from European Mathematical Society. The mini-program consisted of a 4-day conference (January 5–8) and a series of mini-courses, given by the conference speakers before and after the meeting. The focus of the mini-program was variety of concentration phenomena that occur in partial differential equations as well as general theory of concentration analysis. The choice of venue reflects extensive use of concentration methods in Indian mathematics (Adimurthi, Prashanth, Srikanth, Sandeep, Yadava) and the significant role of Indian mathematicians in the global mathematical community. This collection of papers consists mostly of survey articles by conference speakers, representing several research trends in the area.

Concentration analysis is a discipline that straddles partial differential equations, function space theory, geometric analysis, harmonic analysis and functional analysis. The main objective of concentration analysis is to provide, in settings without a priori available compactness, a manageable structural description for the functional sequences intended to approximate solutions of partial differential equations. Use of concentration arguments became an important approach to finding solutions of PDE in the early 1980's due to the works of Uhlenbeck, Brezis, Coron, Lieb, Nirenberg, Aubin, Struwe, and P.-L. Lions. The original method dealt with convergence issues for sequences in Sobolev spaces, whose restrictions to arbitrarily small neighborhoods of certain sets in $\mathbb{R}^N$ (*concentration sets*) retained positive energies preventing compactness, and, conversely, compactness of such sequences followed from the absence of concentration. Perhaps the most famous problem at that time, where the concentration phenomena played the crucial role, was the prescribed scalar curvature (or Yamabe) problem. In a 1984 paper on *global compactness* Michael Struwe gave a representation of bounded Palais–Smale sequences for semilinear elliptic functionals in the form of a finite sum of *elementary*

[1](http://www.icts.res.in/program/ccpd2012)

concentrations, that is, sequences of the form $g_k w$, where (g_k) is a sequence of rescalings (actions of dilations and translations of the form $u \mapsto u(\lambda(\cdot - y))$, $\lambda > 0$, $y \in \mathbb{R}^N$, normalized in the corresponding Sobolev norm) and w, often called the *blowup profile*, satisfies an asymptotic scale-invariant equation, and is typically the famous "standard bubble", also known as instanton or Talenti–Bliss solution. Struwe's global compactness, as it was shown later by Sergio Solimini (Ann. Inst. Henri Poincaré 1995), is a structure common to all sequences bounded in the homogeneous Sobolev space $\mathcal{D}^{1,p}(\mathbb{R}^N)$, $N > p$, once one allows a countable number of elementary concentrations and a remainder vanishing in $L^{\frac{pN}{N-p}}$. Solimini's profile decomposition was independently reproduced, three years later (and with a weaker remainder) the papers of Patrick Gérard ($p = 2$) and Stephane Jaffard (general p), who used the wavelet basis $\Gamma\psi$ generated by actions of a suitable subset Γ of the rescaling group on the "mother wavelet" ψ to form the concentration profiles by considering suitable finite-dimensional spans of the wavelets. The rediscovery of profile decomposition by the "wavelet school" gained substantial circulation and stimulated several important applications to evolution equations (Keraani, Kenig & Merle, Gallagher). A 2003 paper of Schindler and Tintarev generalized the Solimini's result to the abstract Hilbert space H, allowing elementary concentrations $g_k w$ to be formed by a general set D of unitary operators on H. This allowed the inclusion of general *gauge sets* into the framework of concentration analysis such as actions of the conformal groups of Riemannian manifolds, and most recently, the non-standard dilations that form concentrations in the limiting Sobolev dimension (i.e., Moser–Trudinger) imbeddings. The meeting included two communications on profile decompositions for the Moser–Trudinger case.

We refer the reader to a popular description of the topic in a blog entry by Terence Tao written on profile decompositions[2]. To put it concisely, profile decompositions, given by the abstract concentration analysis, involve a D-weakly vanishing remainder sequence r_k, that is a sequence convergent weakly to zero under actions of an arbitrary gauge sequence, $\forall\{g_k\} \subset D, \quad g_k r_k \rightharpoonup 0$ in a given Banach space X. For meaningful applications, a local metrization of such convergence is required, or, in other words, one needs to find a useful Banach space $Y \hookleftarrow X$ (e.g., $Y = L^p$) in which D-weakly vanishing sequences vanish in norm. This gives rise to the notion of D-cocompact imbeddings, that give the analytic grounds for abstract profile decompositions. Different proofs of cocompactness of Sobolev imbeddings are found in papers of Lieb (for subcritical Sobolev imbeddings and translations only) Lions, Solimini and Gerard, and in a monograph of Fieseler and Tintarev. Cocompactness of the Moser–Trudinger imbedding $W_0^{1,2}(\Omega) \hookrightarrow \exp L^2(\Omega)$ is verified in the recent work of Adimurthi–Tintarev as well as in the subsequent work of Bahouri–Majdoub–Masmoudi. Cocompactness, relative to standard rescalings (under the name Assumption 1, which describes a formally stronger, but in many cases equivalent, property), for a large scope of Sobolev-type imbedding involving,

[2]http://terrytao.wordpress.com/2008/11/05/concentration-compactness-and-the-profile-decomposition/

in particular, Triebel–Lizorkin and Besov spaces, has been recently established by Bahouri, Cohen and Koch. Cocompactness in Strichartz imbeddings for dispersive equations, that regulates loss of compactness in the mass-critical case (the energy-critical case is extension of concentration analysis in the Sobolev space) requires engaging a larger group of transformations. In particular, for the Schrödinger equation, this group is a product group of space shifts, dilations and shifts in the Fourier variable.

Profile decompositions of Solimini's type were used initially to establish the compactness of sequences constructed to approximate solutions of nonlinear elliptic problems. If the structure of the equation prevents elementary concentrations from occurring, one concludes compactness. On the other hand, in elliptic problems with a small parameter ϵ, a profile decomposition sequence of Struwe–Solimini type, using ϵ as a concentration parameter, is often a very efficient initial approximation for a solution. Furthermore, it is possible, for positive solutions of semilinear elliptic equations, to associate location of the concentration point with geometry of the problem, In many such problems concentration occurs at the critical points of the Robin function of the domain. For construction of sign-changing solutions, profile decompositions of Struwe–Solimini type, involving profiles (typically "standard bubbles") of alternating signs, provide natural initial approximations, which allow usage of the Lyapunov–Schmidt method to reduce finding the solution to a finite-dimensional problem. As this takes the study beyond the usual energy spaces, this approach allows the construction of solutions for problems with supercritical nonlinearities. Another, similar, use of concentrating approximations is in variational problems of Bahri–Coron type, where topology of the domain affects the geometry of the energy functional, and non-contractive subsets of level sets can be constructed out of families of concentrating profiles, typically involving the standard bubbles. Separation of distinct terms in the profile decompositions is not always spatial, by different concentration cores, but may also occur at the same (or convergent to the same) points, whenever the rates of concentration remain incomparably different. Such "bubble towers" have been also used as initial approximations of solutions.

Blowup analysis of partial differential equations is not limited to solutions of finite energy, that is, to analysis of convergence for sequences bounded in some particular norm. In many cases rescaled limits of sequences of solutions or of approximate solutions satisfy an equation for which a suitable Liouville theorem guarantees that only a trivial solution exists, which in turn provides grounds for existence results. Existence and asymptotics results based on the blowup analysis of PDE at points of maximum has been done in recent years by Adimurthi, Druet, Hebey, Yanyan Li, Robert, Struwe and Juncheng Wei. The conference, furthermore, included some communications on various aspects of non-compact PDE problems that provide basis for future concentration analysis, such as elliptic problems on the hyperbolic space and inequalities of Caffarelli–Kohn–Nirenberg type. Lyapunov–Schmidt method or the finite-dimensional reduction is closely adjacent to concentration analysis. The abstract problem is to find the critical points of a perturbed functional $F_\epsilon = F + \epsilon G$ defined on a Hilbert space for small enough values of ϵ, under the assumption that the critical points of F forms a finite-dimensional manifold Z and the second derivative of F at a point on this manifold is Fredhom of index zero. Then the critical points of F_ϵ can be found by finding the critical points of a functional on the finite-dimensional space Z. The method has its origin in the work of Melnikov in dynamical systems and has been successfully used later to tackle many problems in Hamiltonian system, Yang–Mills equation, perturbations of critical elliptic equations,etc. In recent years this method has been used to study problems which are not in the perturbative form, like proving the existence of sign changing solutions of critical elliptic problems which does not depend on a parameter ϵ. These problems are handled by introducing the number of bubbles as a parameter and then using the Lyapunov–Schmidt method. The conference had a few interesting lectures in these directions on topics like De Giorgi conjecture in higher dimension, Paneitz–Branson operator, singular mean-field equations etc.

Adimurthi and Kunnath Sandeep
TIFR – CAM
P.B. 6503 Sharadanagar
Bangalore 560065, India
e-mail: aditi@math.tiftbng.res.in
sandeep@math.tiftbng.res.in

Ian Schindler
University of Toulouse 1 – Capitole
Manufacture des Tabacs
21, Allée de Brienne
F-31000 Toulouse, France
e-mail: ischndl@univ-tlse1.fr

Cyril Tintarev
Upppsala University
P.O. Box 480
SE-751 06 Uppsala, Sweden
e-mail: tintarev@math.uu.se

Concentration Analysis and Applications to PDE
Trends in Mathematics, 1–15

On the Elements Involved in the Lack of Compactness in Critical Sobolev Embedding

Hajer Bahouri

Abstract. The goal of this paper is to emphasize ideas coming from microlocal analysis to refine the study of lack of compactness in critical Sobolev embedding. In particular, we shall highlight that the elements involving in the characterization of the defect of compactness of critical Sobolev embedding in Lebesgue spaces in [13] are of different kinds than the ones intervening in the study of the critical Sobolev embedding of $H^1(\mathbb{R}^2)$ into the Orlicz space $\mathcal{L}(\mathbb{R}^2)$ in [9]; although in the two frameworks the obstruction to compactness is expressed in the same manner in terms of defect measures. This observation is based on the characterization of these embedding by means of asymptotic decompositions and puts in light the relevance of microlocal analysis behind this approach.

Mathematics Subject Classification (2010). 35A25.

Keywords. Lack of compactness, Orlicz space, frequency localization.

1. Introduction

In [21] and [22], P.-L. Lions investigated the lack of compactness in critical Sobolev embedding both in Lebesgue spaces and in the Orlicz space. These pioneering works have been at the origin of several works for different critical Sobolev embedding through several angles. For instance, in [12] the lack of compactness in the critical Sobolev embedding

$$\dot{H}^s(\mathbb{R}^d) \hookrightarrow L^p(\mathbb{R}^d) \tag{1}$$

in the case where $d \geq 3$ with $0 \leq s < d/2$ and $p = 2d/(d-2s)$ is described in terms of microlocal defect measures and in [13], it is characterized by means of profiles. Later this question is treated in [16] for the more general case of Riesz potential spaces, and recently in [4] the wavelet-based profile decomposition introduced by S. Jaffard is revisited in order to treat a larger range of examples of critical embedding of functions spaces

$$X \hookrightarrow Y \tag{2}$$

including Sobolev, Besov, Triebel–Lizorkin, Lorentz, Hölder and BMO spaces. (One can consult [3] and the references therein for an introduction to these spaces.)

In [7, 8, 9], the lack of compactness of the critical Sobolev embedding

$$H^1(\mathbb{R}^2) \hookrightarrow \mathcal{L}(\mathbb{R}^2), \tag{3}$$

where $\mathcal{L}$ denotes the Orlicz space associated to the function $\phi(s) = \mathrm{e}^{s^2} - 1$ (introduced in Definition 1 below) was characterized in terms of an asymptotic decomposition by means of generalization of the example by Moser derived in [24] (see (19) for a precise expression of this example).

Other studies have been conducted in various works supplying us with a large amount of information about solutions of nonlinear partial differential equations, both in the elliptic frame or the evolution frame. Among others, one can mention [2, 5, 6, 10, 11, 15, 17, 18, 19, 23, 28, 29, 30].

The several works devoted to this subject relate the lack of compactness in critical Sobolev embedding to a concentration effect. In what follows, we shall highlight that the phenomenon responsible for the defect of compactness of $\dot{H}^s(\mathbb{R}^d) \hookrightarrow L^p(\mathbb{R}^d)$ and more generally of the embedding (2) is strikingly different from the one arising in the lack of compactness of the embedding of $H^1(\mathbb{R}^2)$ into the Orlicz space $\mathcal{L}(\mathbb{R}^2)$; while in the two cases the obstacle to compactness is expressed in the same manner in terms of defect measures. This fact is due to the characterization of these embedding by means of asymptotically orthogonal decompositions and reveals the importance of microlocal analysis behind this approach.

It turns out that the profiles involved in the decomposition derived by P. Gérard in [13] (or by S. Jaffard in [16] and by H. Bahouri, A. Cohen and G. Kock in [4] and the references therein) are not the right concept to describe the lack of compactness of $H^1(\mathbb{R}^2)$ into $\mathcal{L}(\mathbb{R}^2)$. In fact, in [8] the lack of compactness of the critical Sobolev embedding (3) was characterized by means of an asymptotically orthogonal decomposition in terms of elementary concentrations under the form:

$$g_n(x) := \sqrt{\frac{\alpha_n}{2\pi}}\, \psi\Big(\frac{-\log|x - x_n|}{\alpha_n}\Big), \tag{4}$$

where (α_n) is a sequence of positive real numbers going to infinity, (x_n) is a convergent sequence of points in $\mathbb{R}^2$ and ψ belongs to the set $\mathcal{P}$ introduced in Definition 2.

In [13], P. Gérard reduced the lack of compactness of the critical Sobolev embedding (1) to elementary concentrations taking the form:

$$f_n(x) := \frac{1}{(h_n)^{\frac{d}{p}}} \varphi\left(\frac{x - x_n}{h_n}\right), \tag{5}$$

where (h_n) is a sequence of positive real numbers, (x_n) is a convergent sequence of points in $\mathbb{R}^d$ and φ belongs to $\dot{H}^s(\mathbb{R}^d)$. Note that the defect of compactness in the more general framework of abstract critical Sobolev embedding (2) investigated in [4] is due, up to scale invariance, to elements of the same kind.

As we shall prove in the sequel, the elementary concentrations g_n given by (4) are strikingly different from the ones given by (5). In one hand with the vocabulary of [12] (see also Definition 4 in this paper), the profiles f_n defined above are h_n-oscillating while the elements g_n introduced in (4) are unrelated to any frequency. On the other hand as can be seen in [4, 7, 8, 9, 12, 16], the approach adopted to extract the elementary concentrations (4) is based on capacity arguments and outlines that within the framework of the Orlicz space the lack of compactness is due to big values taken by the sequences in consistent sets of size; whereas the process in the characterization of the lack of compactness (1) and more generally (2) uses in a crucial way the existence of oscillating components.

The structure of the paper is the following. First in Section 2, we present a short review of characterization of lack of compactness in critical Sobolev embedding, recalling two types of description results by means of asymptotically orthogonal decompositions. Section 3 is devoted to the statement and the proof of the results. Finally, we mention that C will be used to denote a constant which may vary from line to line and for simplicity, we shall also still denote by (u_n) any subsequence of (u_n).

2. Development on lack of compactness in critical Sobolev embedding

2.1. Critical Sobolev embedding in Orlicz space

It is known that $H^1(\mathbb{R}^2)$ embeds into the Orlicz space $\mathcal{L}(\mathbb{R}^2)$ and that

$$\|u\|_{\mathcal{L}} \leq \frac{1}{\sqrt{4\pi}}\|u\|_{H^1}, \tag{6}$$

where $\mathcal{L}$ denotes the Orlicz space associated to the function $\phi(s) = \mathrm{e}^{s^2} - 1$ (see Definition 1 below). This embedding derives immediately from the following Trudinger–Moser type inequalities the proof of which can be found in [1, 24, 27]:

Proposition 1.

$$\sup_{\|u\|_{H^1}\leq 1} \int_{\mathbb{R}^2} \left(\mathrm{e}^{4\pi|u(x)|^2} - 1\right) dx := \kappa < \infty. \tag{7}$$

Let us recall that generally, if $\phi : \mathbb{R}^+ \to \mathbb{R}^+$ is a convex increasing function such that

$$\phi(0) = 0 = \lim_{s\to 0^+} \phi(s) \text{ and } \lim_{s\to\infty} \phi(s) = \infty,$$

then L^ϕ the Orlicz space on $\mathbb{R}^d$ associated to the function ϕ is defined as follows (for a complete presentation on Orlicz spaces, we refer the reader to [26] and the references therein):

Definition 1. We say that a measurable function $u : \mathbb{R}^d \to \mathbb{C}$ belongs to L^ϕ if there exists $\lambda > 0$ such that

$$\int_{\mathbb{R}^d} \phi\left(\frac{|u(x)|}{\lambda}\right) dx < \infty$$

and we denote

$$\|u\|_{L^\phi} = \inf \left\{ \lambda > 0, \quad \int_{\mathbb{R}^d} \phi\left(\frac{|u(x)|}{\lambda}\right) dx \leq 1 \right\}. \tag{8}$$

Remark 1. In the sequel we shall fix $d = 2$, $\phi(s) = \mathrm{e}^{s^2} - 1$ and denote the Orlicz space L^ϕ by $\mathcal{L}$ endowed with the norm $\|\cdot\|_{\mathcal{L}}$ where the number 1 is replaced by the constant κ appearing in Identity (7).

Before coming to the statement of the characterization of the lack of compactness of the critical Sobolev embedding (3), let us introduce some objects.

Definition 2. We shall designate by scale any sequence $\underline{\alpha} := (\alpha_n)$ of positive real numbers going to infinity, by core any sequence $\underline{x} := (x_n)$ of points in $\mathbb{R}^2$ and by profile any function ψ belonging to the set

$$\mathcal{P} := \left\{ \psi \in L^2(\mathbb{R}, \mathrm{e}^{-2s} ds); \quad \psi' \in L^2(\mathbb{R}) \text{ and } \psi_{|]-\infty,0]} = 0 \right\}.$$

Given two scales $\underline{\alpha}$, $\underline{\tilde{\alpha}}$, two cores $\underline{x}$, $\underline{\tilde{x}}$ and two profiles ψ, $\tilde{\psi}$, we say that the triplets $(\underline{\alpha}, \underline{x}, \psi)$ and $(\underline{\tilde{\alpha}}, \underline{\tilde{x}}, \tilde{\psi})$ are orthogonal if

either

$$\Big| \log\left(\tilde{\alpha}_n/\alpha_n\right) \Big| \longrightarrow \infty\,, \tag{9}$$

or

$$\tilde{\alpha}_n = \alpha_n \quad \text{and} \quad \Big(-\frac{\log|x_n - \tilde{x}_n|}{\alpha_n} \longrightarrow a \geq 0 \text{ with } \psi \text{ or } \tilde{\psi} \text{ null for } s < a \Big). \tag{10}$$

Remark 2. Starting from the fact that $\psi(s) = \int_0^s \psi'(\tau)\, d\tau$, one can easily prove that

$$\max_{s>0} \frac{|\psi(s)|}{\sqrt{s}} \leq \|\psi'\|_{L^2}. \tag{11}$$

This leads us to distinguish two cases of elementary concentrations (4): a first case where the inequality in (11) is strict and a second case where this inequality is an equality. The second case corresponds to the case where the whole mass concentrates and reduces to the example derived by Moser in [24] (for a proof of this fact, see [7]).

The asymptotically orthogonal decomposition derived in [8] is formulated in the following terms:

Theorem 1. *Let (u_n) be a bounded sequence in $H^1(\mathbb{R}^2)$ such that*

$$u_n \rightharpoonup 0, \tag{12}$$

$$\limsup_{n\to\infty} \|u_n\|_{\mathcal{L}} = A_0 > 0 \tag{13}$$

and

$$\lim_{R\to\infty}\ \limsup_{n\to\infty}\ \|u_n\|_{\mathcal{L}(|x|>R)}=0. \tag{14}$$

Then, there exist a sequence of scales $(\underline{\alpha}^{(j)})$, a sequence of cores $(\underline{x}^{(j)})$ and a sequence of profiles $(\psi^{(j)})$ such that the triplets $(\underline{\alpha}^{(j)}, \underline{x}^{(j)}, \psi^{(j)})$ are pairwise orthogonal and, up to a subsequence extraction, we have for all $\ell \geq 1$,

$$u_n(x)=\sum_{j=1}^{\ell}\sqrt{\frac{\alpha_n^{(j)}}{2\pi}}\,\psi^{(j)}\left(\frac{-\log|x-x_n^{(j)}|}{\alpha_n^{(j)}}\right)+\mathrm{r}_n^{(\ell)}(x),\quad \limsup_{n\to\infty}\ \|\mathrm{r}_n^{(\ell)}\|_{\mathcal{L}}\stackrel{\ell\to\infty}{\longrightarrow}0. \tag{15}$$

Remark 3.

- It was shown in [7] that

$$\lim_{n\to\infty}\|g_n^{(j)}\|_{\mathcal{L}}=\frac{1}{\sqrt{4\pi}}\max_{s>0}\frac{|\psi^{(j)}(s)|}{\sqrt{s}}\quad\text{and} \tag{16}$$

$$|\nabla g_n^{(j)}(x)|^2\,dx \rightharpoonup \|\psi^{(j)\prime}\|_{L^2}^2\,\delta_{x_0},\quad\text{in}\quad \mathcal{D}'(\mathbb{R}^2), \tag{17}$$

when the sequence (x_n) converges towards x_0 and where

$$g_n^{(j)}=\sqrt{\frac{\alpha_n^{(j)}}{2\pi}}\,\psi^{(j)}\left(\frac{-\log|x-x_n^{(j)}|}{\alpha_n^{(j)}}\right).$$

- It is easy to verify that

$$\|\nabla g_n^{(j)}\|_{L^2(\mathbb{R}^2)}=\|\psi^{(j)\prime}\|_{L^2([0,+\infty[)} \tag{18}$$

- It should be noted that the elementary concentrations involving in (15) are generalization of the example derived by Moser in [24]. This example defined as follows:

$$f_{\alpha_n}(x)=\begin{cases}0 & \text{if}\quad |x|\geq 1,\\ -\frac{\log|x|}{\sqrt{2\alpha_n\pi}} & \text{if}\quad \mathrm{e}^{-\alpha_n}\leq|x|\leq 1,\\ \sqrt{\frac{\alpha_n}{2\pi}} & \text{if}\quad |x|\leq \mathrm{e}^{-\alpha_n},\end{cases} \tag{19}$$

can be written as

$$f_{\alpha_n}(x)=\sqrt{\frac{\alpha_n}{2\pi}}\,\mathbf{L}\Big(\frac{-\log|x|}{\alpha_n}\Big),$$

where

$$\mathbf{L}(s)=\begin{cases}0 & \text{if}\quad s\leq 0,\\ s & \text{if}\quad 0\leq s\leq 1,\\ 1 & \text{if}\quad s\geq 1\end{cases} \tag{20}$$

and satisfies $\|f_{\alpha_n}\|_{\mathcal{L}}\to\frac{1}{\sqrt{4\pi}}\max_{s>0}\frac{|\mathbf{L}(s)|}{\sqrt{s}}=\frac{1}{\sqrt{4\pi}}$, as $\alpha_n\to\infty$.

- The heart of the matter relies on the fact that contrary to elementary concentrations given by (5), the frequency of such sequences are spread. More precisely according to Definition 4, (∇f_{α_n}) is unrelated to any frequency. This will be the subject of Theorem 3.

2.2. Critical Sobolev embedding in Lebesgue spaces

It is well known that $\dot{H}^s(\mathbb{R}^d)$ embeds into the Lebesgue space $L^p(\mathbb{R}^d)$, in the case where $d \geq 3$ with $0 \leq s < d/2$ and $p = 2d/(d-2s)$ and that this embedding is non compact. In order to recall the result of P. Gérard ([12]) in a clear way, let us first introduce some related definitions.

Definition 3. A scale is a sequence $h := (h_n)$ of positive real numbers going to 0, 1 or infinity, a core is a sequence $x := (x_n)$ of points in $\mathbb{R}^d$ and a profile is a function $\varphi \in \dot{H}^s(\mathbb{R}^d)$. Given two scales h, $\tilde{h}$ and two cores x, $\tilde{x}$, we say that the couples (h, x) and $(\tilde{h}, \tilde{x})$ are orthogonal if

$$\text{either} \quad \left|\log\left(\tilde{h}_n/h_n\right)\right| \overset{n\to\infty}{\longrightarrow} \infty \quad \text{or} \quad \left(\tilde{h}_n = h_n \quad \text{and} \quad \frac{|x_n - \tilde{x}_n|}{h_n} \overset{n\to\infty}{\longrightarrow} \infty\right). \tag{21}$$

The profile decomposition theorem in that framework states as follows:

Theorem 2. *Let (u_n) be a bounded sequence in $\dot{H}^s(\mathbb{R}^d)$. Then, there exist a sequence of scales $(h^{(j)})$, a sequence of cores $(x^{(j)})$ and a sequence of profiles $(\varphi^{(j)})$ such that the couples $(h^{(j)}, x^{(j)})$ are pairwise orthogonal and, up to a subsequence extraction, we have for all $\ell \geq 1$,*

$$u_n(x) = \sum_{j=1}^{\ell} \left(\frac{1}{h_n^{(j)}}\right)^{\frac{d}{p}} \varphi^{(j)}\left(\frac{x - x_n^{(j)}}{h_n^{(j)}}\right) + \mathrm{r}_n^{(\ell)}(x), \quad \limsup_{n\to\infty} \|\mathrm{r}_n^{(\ell)}\|_{L^p(\mathbb{R}^d)} \overset{\ell\to\infty}{\longrightarrow} 0. \tag{22}$$

Remark 4. As it is mentioned above and contrary to $\nabla g_n^{(j)}$, the sequence $|D|^s f_n^{(j)}$ is h-oscillating according to Definition 4, where $f_n^{(j)} = \left(\frac{1}{h_n^{(j)}}\right)^{\frac{d}{p}} \varphi^{(j)}\left(\frac{x - x_n^{(j)}}{h_n^{(j)}}\right)$ and $\widehat{|D|^s f_n^{(j)}}(\xi) = |\xi|^s \widehat{f_n^{(j)}}(\xi)$. Indeed, straightforward computations lead to

$$\int_{h_n|\xi|\leq \frac{1}{R}} |\widehat{|D|^s f_n^{(j)}}(\xi)|^2 \, d\xi = \int_{|\eta|\leq \frac{1}{R}} |\eta|^{2s} |\widehat{\varphi^{(j)}}(\eta)|^2 \, d\eta .$$

As well

$$\int_{h_n|\xi|\geq R} |\widehat{|D|^s f_n^{(j)}}(\xi)|^2 \, d\xi = \int_{|\eta|\geq R} |\eta|^{2s} |\widehat{\varphi^{(j)}}(\eta)|^2 \, d\eta ,$$

which ensures the result.

3. Statement and proof of the results

3.1. Presentation of the results

Our purpose is to show that the elementary concentrations

$$g_n^{(j)} = \sqrt{\frac{\alpha_n^{(j)}}{2\pi}}\, \psi^{(j)} \left(\frac{-\log|x - x_n^{(j)}|}{\alpha_n^{(j)}} \right)$$

involving in the asymptotic decomposition (15) are completely different from the profiles characterizing the lack of compactness of the Sobolev embedding investigated in [4], [13] and [16], although in the two frameworks the obstruction to compactness is expressed in the same manner in terms of defect measures. Before coming to the statement of the result, let us recall the notions of being oscillating with respect to a scale and of being unrelated to any scale introduced by P. Gérard in [13].

Definition 4. Let $f := (f_n)$ be a bounded sequence in $L^2(\mathbb{R}^d)$ and $h := (h_n)$ be a sequence of positive reals.

- The sequence f is said h-oscillating if

$$\limsup_{n\to\infty} \left(\int_{h_n|\xi| \leq \frac{1}{R}} |\widehat{f_n}(\xi)|^2 \, d\xi + \int_{h_n|\xi|\geq R} |\widehat{f_n}(\xi)|^2 \, d\xi \right) \overset{R\to\infty}{\longrightarrow} 0 . \tag{23}$$

- The sequence f is said unrelated to the scale h if for any reals $b > a > 0$

$$\int_{a \leq h_n|\xi| \leq b} |\widehat{f_n}(\xi)|^2 \, d\xi \overset{n\to\infty}{\longrightarrow} 0 . \tag{24}$$

Our main result states as follows:

Theorem 3. *Let (α_n) be a scale, (x_n) be a core and ψ a profile in the sense of Definition 2, then the generalizing example by Moser*

$$g_n(x) = \sqrt{\frac{\alpha_n}{2\pi}}\, \psi \left(\frac{-\log|x - x_n|}{\alpha_n} \right)$$

is unrelated to any scale, in the sense that for any reals $0 < a < b$ and any sequence (h_n) of nonnegative real numbers

$$\int_{a<h_n|\xi|<b} |\widehat{\nabla g_n}(\xi)|^2 \, d\xi \longrightarrow 0, \quad \text{as} \quad n \to \infty. \tag{25}$$

As we shall see later on, in the framework of the critical Sobolev embedding $H^1(\mathbb{R}^2) \hookrightarrow \mathcal{L}(\mathbb{R}^2)$, it is more relevant to introduce the following notions:

Definition 5. Let $h := (h_n)$ be a bounded sequence in $L^2(\mathbb{R}^2)$ and $\alpha := (\alpha_n)$ be a sequence of positive reals.

- The sequence h is said $(\log -\alpha)$-oscillating if

$$\limsup_{n\to\infty} \left(\int_{\frac{1}{\alpha_n}(\log|\xi|) \leq \frac{1}{R}} |\widehat{h_n}(\xi)|^2 \, d\xi + \int_{\frac{1}{\alpha_n}(\log|\xi|)\geq R} |\widehat{h_n}(\xi)|^2 \, d\xi \right) \overset{R\to\infty}{\longrightarrow} 0 . \tag{26}$$

- The sequence h is said log-unrelated to the scale α if for any reals $b > a > 0$

$$\int_{a\leq \frac{1}{\alpha_n}(\log|\xi|)\leq b} |\widehat{h_n}(\xi)|^2\, d\xi \overset{n\to\infty}{\longrightarrow} 0\,. \tag{27}$$

In fact, the sequence (∇g_n) satisfies the following property:

Proposition 2. *With the vocabulary of Definition* 5, *the sequence* (∇g_n) *is* $(\log - \alpha)$-*oscillating.*

Remark 5.

- It was emphasized in [13] that the characteristic of being unrelated to any scale in the sense of (24) is measured using the Besov norm $\dot{B}^0_{2,\infty}$. Then in view of (25), we have

$$\|g_n\|_{\dot{B}^1_{2,\infty}(\mathbb{R}^2)} \longrightarrow 0, \quad \text{as} \quad n\to\infty. \tag{28}$$

- In the case where $g_n = f_{\alpha_n}$ with f_{α_n} the example by Moser, we can already observe that

$$\widehat{f_{\alpha_n}}(\xi) = \sqrt{2\pi\,\alpha_n}\int_0^\infty J_0\Big(r|\xi|\Big)\,\mathbf{L}\Big(\frac{-\log r}{\alpha_n}\Big)\,rdr,$$

where J_0 is the Bessel function solution of the differential equation

$$xy'' + y' + xy = 0, \tag{29}$$

that can be written under the form

$$(xy')' + xy = 0. \tag{30}$$

Performing the change of variables $r = \mathrm{e}^{-\alpha_n s}$, we get taking account of the fact that $\mathbf{L}_{|]-\infty,0]} = 0$

$$\widehat{f_{\alpha_n}}(\xi) = \alpha_n\sqrt{2\pi\,\alpha_n}\int_0^\infty J_0(\mathrm{e}^{-\alpha_n s}|\xi|)\,\mathbf{L}(s)\,\mathrm{e}^{-2\,\alpha_n s}ds.$$

Now setting $t = \mathrm{e}^{-\alpha_n s}|\xi|$, we get in view of (30) for $\xi \neq 0$

$$\begin{aligned}\mathrm{e}^{-\alpha_n s}\,|\xi|\,J_0(\mathrm{e}^{-\alpha_n s}|\xi|) &= -\partial_t\Big(\mathrm{e}^{-\alpha_n s}\,|\xi|\,J_0'(\mathrm{e}^{-\alpha_n s}|\xi|)\Big)\\ &= \frac{\partial_s\Big(\mathrm{e}^{-\alpha_n s}\,|\xi|\,J_0'(\mathrm{e}^{-\alpha_n s}|\xi|)\Big)}{\alpha_n\,\mathrm{e}^{-\alpha_n s}\,|\xi|}\cdot\end{aligned}$$

Integrating by parts, it comes according to (20)

$$\widehat{f_{\alpha_n}}(\xi) = -\frac{1}{|\xi|}\sqrt{2\pi\,\alpha_n}\int_0^1 \mathrm{e}^{-\alpha_n s}\,J_0'(\mathrm{e}^{-\alpha_n s}|\xi|)\,ds\cdot$$

Finally, the change of variables $u = \mathrm{e}^{-\alpha_n s}|\xi|$ leads for $\xi\neq 0$ to

$$\widehat{f_{\alpha_n}}(\xi) = \frac{1}{|\xi|^2}\sqrt{\frac{2\pi}{\alpha_n}}\Big(J_0(\mathrm{e}^{-\alpha_n}|\xi|) - J_0(|\xi|)\Big)\cdot \tag{31}$$

In the following subsection, we shall take advantage of Formula (31) to express in a useful way the example by Moser.

3.2. Specific results on the example by Moser

First, let us point out that the scales of f_{α_n} are spread in the ball of radius e^{α_n}. More precisely, we have the following result:

Proposition 3. *The sequence* (∇f_{α_n}) *satisfies*

$$\int_{|\xi|\geq \mathrm{e}^{\alpha_n}} |\widehat{\nabla f_{\alpha_n}}(\xi)|^2\, d\xi \longrightarrow 0, \quad as \quad n\to\infty. \tag{32}$$

Proof. In view of (31), we have

$$\int_{|\xi|\geq \mathrm{e}^{\alpha_n}} |\widehat{\nabla f_{\alpha_n}}(\xi)|^2\, d\xi = \frac{(2\pi)^2}{\alpha_n}\int_{\mathrm{e}^{\alpha_n}}^{\infty} \Big|J_0\big(\mathrm{e}^{-\alpha_n} r\big) - J_0(r)\Big|^2 \frac{dr}{r}.$$

Before entering into the details, let us recall that as x tends to infinity (see for instance [20, 25]):

$$J_0(x) = \sqrt{\frac{2}{\pi x}}\cos\Big(x-\frac{\pi}{4}\Big) + O(x^{-\frac{3}{2}}), \tag{33}$$

and

$$J_0'(x) = -\sqrt{\frac{2}{\pi x}}\sin\Big(x-\frac{\pi}{4}\Big) + O(x^{-\frac{3}{2}}). \tag{34}$$

We deduce that for $r\geq \mathrm{e}^{\alpha_n}$

$$J_0\big(\mathrm{e}^{-\alpha_n} r\big) = \sqrt{\frac{2\,\mathrm{e}^{\alpha_n}}{\pi r}}\cos\Big(\mathrm{e}^{-\alpha_n}r-\frac{\pi}{4}\Big) + O(\mathrm{e}^{\frac{3}{2}\alpha_n}\, r^{-\frac{3}{2}}),$$

and

$$J_0(r) = \sqrt{\frac{2}{\pi r}}\cos\Big(r-\frac{\pi}{4}\Big) + O(r^{-\frac{3}{2}}).$$

Therefore making use of the asymptotic development, we get

$$\int_{|\xi|\geq \mathrm{e}^{\alpha_n}} |\widehat{\nabla f_{\alpha_n}}(\xi)|^2\, d\xi \lesssim I_1 + I_2,$$

with

$$I_1 := \frac{16\,\pi}{\alpha_n}\int_{\mathrm{e}^{\alpha_n}}^{\infty}\Big|\sqrt{\frac{\mathrm{e}^{\alpha_n}}{r}}\cos\Big(\mathrm{e}^{-\alpha_n}r-\frac{\pi}{4}\Big) - \frac{1}{\sqrt{r}}\cos\Big(r-\frac{\pi}{4}\Big)\Big|^2\frac{dr}{r},$$

and

$$I_2 := \frac{C\,\mathrm{e}^{3\,\alpha_n}}{\alpha_n}\int_{\mathrm{e}^{\alpha_n}}^{\infty}\frac{dr}{r^4}.$$

It is obvious that $I_2 \leq \frac{C}{\alpha_n} \longrightarrow 0$, as $n\to\infty$. Also, we have

$$I_1 \leq \frac{C\,\mathrm{e}^{\alpha_n}}{\alpha_n}\int_{\mathrm{e}^{\alpha_n}}^{\infty}\frac{dr}{r^2} = \frac{C}{\alpha_n} \longrightarrow 0, \quad \text{as} \quad n\to\infty,$$

which ends the proof of the proposition. □

Remark 6.

- Now, let us observe that for any fixed M
$$\int_{0\leq|\xi|\leq M} |\widehat{\nabla f_{\alpha_n}}(\xi)|^2\, d\xi \leq M^2 \|f_{\alpha_n}\|^2_{L^2} \stackrel{n\to\infty}{\longrightarrow} 0\cdot \tag{35}$$
- Besides for any fixed M, we have in light of (33)
$$\frac{1}{\alpha_n}\int_M^{e^{\alpha_n}} \big|J_0(r)\big|^2 \frac{dr}{r} \leq \frac{C}{\alpha_n}\int_M^{e^{\alpha_n}} \frac{dr}{r^2} \stackrel{n\to\infty}{\longrightarrow} 0\cdot \tag{36}$$
- Finally let us notice that for any fixed M,
$$\frac{1}{\alpha_n}\int_M^{e^{\alpha_n}} \big|J_0(e^{-\alpha_n} r) - J_0(0)\big|^2 \frac{dr}{r} \stackrel{n\to\infty}{\longrightarrow} 0\cdot \tag{37}$$
Indeed in view of Taylor's formula, we have
$$J_0(e^{-\alpha_n} r) - J_0(0) = e^{-\alpha_n}\, r \int_0^1 J_0'\big(t\, e^{-\alpha_n}\, r\big)\, dt\,,$$
which implies that
$$\frac{1}{\alpha_n}\int_M^{e^{\alpha_n}} \big|J_0(e^{-\alpha_n} r) - J_0(0)\big|^2 \frac{dr}{r} \leq \frac{C\, e^{-2\alpha_n}}{\alpha_n}\int_M^{e^{\alpha_n}} r\, dr \leq \frac{C}{2\,\alpha_n} \stackrel{n\to\infty}{\longrightarrow} 0\cdot$$
The above computations conduct to the following result:

Proposition 4. *The sequence (f_{α_n}) also writes under the form:*
$$f_{\alpha_n}(x) = \frac{1}{(2\pi)^2}\sqrt{\frac{2\pi}{\alpha_n}}\int_{1\leq|\xi|\leq e^{\alpha_n}} e^{i\,x\cdot\xi}\frac{1}{|\xi|^2}\, d\xi + \mathrm{r}_n \tag{38}$$
with $\|\nabla \mathrm{r}_n\|_{L^2} \longrightarrow 0$, *as* $n \to \infty$.

Remark 7.

- According to (38), we can easily prove that the example by Moser ∇f_{α_n} is $(\log -\alpha)$-oscillating. Indeed for $R \geq 1$
$$\int_{\frac{1}{\alpha_n}(\log|\xi|)\geq R} |\widehat{\nabla f_{\alpha_n}}(\xi)|^2\, d\xi \leq \int_{|\xi|\geq e^{\alpha_n}} |\widehat{\nabla f_{\alpha_n}}(\xi)|^2\, d\xi \longrightarrow 0, \quad \text{as} \quad n\to\infty\,,$$
in light of (32). Besides,
$$\int_{\frac{1}{\alpha_n}(\log|\xi|)\leq\frac{1}{R}} |\widehat{\nabla f_{\alpha_n}}(\xi)|^2\, d\xi = \int_{|\xi|\leq 1} |\widehat{\nabla f_{\alpha_n}}(\xi)|^2\, d\xi + \int_{1\leq|\xi|\leq e^{\frac{\alpha_n}{R}}} |\widehat{\nabla f_{\alpha_n}}(\xi)|^2\, d\xi\,.$$
Firstly by (35) $\int_{|\xi|\leq 1} |\widehat{\nabla f_{\alpha_n}}(\xi)|^2\, d\xi \stackrel{n\to\infty}{\longrightarrow} 0$, and secondly thanks to (38)
$$\int_{1\leq|\xi|\leq e^{\frac{\alpha_n}{R}}} |\widehat{\nabla f_{\alpha_n}}(\xi)|^2\, d\xi \leq \frac{C}{\alpha_n}\int_1^{e^{\frac{\alpha_n}{R}}} \frac{dr}{r} + \circ(1) \leq \frac{C}{R} + \circ(1)\,,$$
where $\circ(1)$ denotes a sequence which tends to 0 as n goes to infinity. This ends the proof of the result.

- If we denote

$$\widetilde{f}_{\alpha_n}(x) = \frac{1}{(2\pi)^2}\sqrt{\frac{2\pi}{\alpha_n}}\int_{1\leq|\xi|\leq e^{\alpha_n}} e^{i\,x\cdot\xi}\,\frac{1}{|\xi|^2}\,d\xi\,,$$

then $\widetilde{f}_{\alpha_n}(0) = \sqrt{\frac{\alpha_n}{2\pi}}$ and for $|x| \leq e^{-\alpha_n}$

$$\widetilde{f}_{\alpha_n}(x) = \sqrt{\frac{\alpha_n}{2\pi}} + \circ(1)\,, \tag{39}$$

where $\circ(1)$ still denotes a sequence which tends to 0 as n goes to infinity. Indeed, for $|x| \leq e^{-\alpha_n}$

$$\begin{aligned}\widetilde{f}_{\alpha_n}(x) &= \frac{1}{\sqrt{2\pi\alpha_n}}\int_1^{e^{\alpha_n}} J_0(|x|\rho)\,\frac{d\rho}{\rho}\\ &= \frac{1}{\sqrt{2\pi\alpha_n}}\int_1^{e^{\alpha_n}} J_0(0)\,\frac{d\rho}{\rho} + \frac{|x|}{\sqrt{2\pi\alpha_n}}\int_1^{e^{\alpha_n}}\Big(\int_0^1 J_0'(t|x|\rho)\,dt\Big)\,d\rho\\ &= \sqrt{\frac{\alpha_n}{2\pi}} + \widetilde{t}_n(x)\,,\end{aligned}$$

with $|\widetilde{t}_n(x)| \leq \frac{|x|e^{\alpha_n}}{\sqrt{2\pi\,\alpha_n}}\sup_{0\leq u\leq 1}|J_0'(u)|$. This ensures the result.
- Taking advantage of (39), we can easily check that $\|\widetilde{f}_{\alpha_n}\|_{\mathcal{L}} \to \frac{1}{\sqrt{4\pi}}$, as $\alpha_n \to \infty$.

3.3. Proofs of Theorem 3 and Proposition 2

First of all, let us observe that for any fixed M

$$\int_{0\leq|\xi|\leq M} |\widehat{\nabla g_n}(\xi)|^2\,d\xi \leq M^2\|g_n\|_{L^2}^2 \overset{n\to\infty}{\longrightarrow} 0. \tag{40}$$

Therefore in what follows, we shall focus on the area $\{\xi,\,|\xi| \geq 1\}$. Now, along the same lines as the proof of (31), we get using the fact that $\psi(0) = 0$ and $|\psi(s)| \leq C\sqrt{s}$ for all $s \geq 0$

$$\begin{aligned}\widehat{g_n}(\xi) &= \frac{1}{|\xi|}\sqrt{2\pi\,\alpha_n}\int_0^\infty \partial_s\Big(e^{-\alpha_n s}\,J_0'(e^{-\alpha_n s}|\xi|)\Big)\,\psi(s)\,ds\\ &= -\frac{1}{|\xi|}\sqrt{2\pi\,\alpha_n}\int_0^\infty e^{-\alpha_n s}\,J_0'(e^{-\alpha_n s}|\xi|)\,\psi'(s)\,ds.\end{aligned} \tag{41}$$

To go to the proof of (25), we shall split $|\xi|\,\widehat{g_n}(\xi)$ into two parts:

$$I := |\xi|\,\widehat{g_n}(\xi) = I_1 + I_2,$$

with

$$I_1 := -\sqrt{2\pi\,\alpha_n}\int_0^{\frac{1}{\alpha_n}\log(|\xi|)} e^{-\alpha_n s}\,J_0'(e^{-\alpha_n s}|\xi|)\,\psi'(s)\,ds,$$

and

$$I_2 := -\sqrt{2\pi\,\alpha_n}\int_{\frac{1}{\alpha_n}\log(|\xi|)}^\infty e^{-\alpha_n s}\,J_0'(e^{-\alpha_n s}|\xi|)\,\psi'(s)\,ds.$$

Let us begin by assuming that $\psi' \in \mathcal{D}(]0,+\infty[)$. It is then obvious that

$$I_2 \leq \frac{C\|\psi'\|_{L^\infty}}{\sqrt{\alpha_n}\,|\xi|}\cdot$$

To estimate I_1, we shall make use of the asymptotic development (34) and write

$$I_1 := J_1 + J_2,$$

with

$$J_1 := 2\sqrt{\alpha_n}\int_0^{\frac{1}{\alpha_n}\log(|\xi|)} \mathrm{e}^{-\alpha_n s}\,\frac{\sin\left(\mathrm{e}^{-\alpha_n s}|\xi| - \frac{\pi}{4}\right)}{\sqrt{\mathrm{e}^{-\alpha_n s}|\xi|}}\,\psi'(s)\,ds,$$

and

$$J_2 := \sqrt{2\pi\,\alpha_n}\int_0^{\frac{1}{\alpha_n}\log(|\xi|)} \mathrm{e}^{-\alpha_n s}\,O((\mathrm{e}^{-\alpha_n s}|\xi|)^{-\frac{3}{2}})\,\psi'(s)\,ds\cdot$$

It is clear that

$$|J_2| \leq \frac{C\|\psi'\|_{L^\infty}}{\sqrt{\alpha_n}|\xi|}\cdot$$

Now to estimate J_1, we integrate by parts. Then, we get

$$J_1 = \frac{2}{\sqrt{\alpha_n}\,|\xi|^{\frac{3}{2}}}\Big(\Big[\cos\left(\mathrm{e}^{-\alpha_n s}|\xi| - \frac{\pi}{4}\right)\mathrm{e}^{\frac{\alpha_n s}{2}}\psi'(s)\Big]_0^{\frac{1}{\alpha_n}\log(|\xi|)}$$
$$- \int_0^{\frac{1}{\alpha_n}\log(|\xi|)} \mathrm{e}^{\frac{\alpha_n s}{2}}\left(\frac{\alpha_n}{2}\psi'(s) + \psi''(s)\right)\cos\left(\mathrm{e}^{-\alpha_n s}|\xi| - \frac{\pi}{4}\right)ds\Big),$$

which implies by straightforward computations that

$$|J_1| \leq \frac{C}{\sqrt{\alpha_n}\,|\xi|}\Big(\|\psi'\|_{L^\infty} + \|\psi''\|_{L^\infty}\Big)\cdot$$

To summarize, we have in the case when $\psi' \in \mathcal{D}(]0,+\infty[)$

$$|\xi||\widehat{g_n}(\xi)| \leq \frac{C_\psi}{\sqrt{\alpha_n}\,|\xi|}\cdot \tag{42}$$

Consequently

$$\int_{a<h_n|\xi|<b} |\widehat{\nabla g_n}(\xi)|^2\,d\xi \leq \frac{C_\psi}{\alpha_n}\int_{\frac{a}{h_n}<\rho<\frac{b}{h_n}} \frac{d\rho}{\rho} \leq \frac{C_\psi}{\alpha_n}\log\left(\frac{b}{a}\right) \overset{n\to\infty}{\longrightarrow} 0.$$

This achieves the proof of the theorem in view of (18).

To end this section, let us prove Proposition 2. Along the same lines as in the proof of Theorem 3, one can reduce to the case where $\psi' \in \mathcal{D}(]0,+\infty[)$. Firstly

$$\int_{\frac{1}{\alpha_n}(\log|\xi|)\leq\frac{1}{R}} |\widehat{\nabla g_n}(\xi)|^2\,d\xi = \int_{|\xi|\leq 1} |\widehat{\nabla g_n}(\xi)|^2\,d\xi + \int_{1\leq|\xi|\leq \mathrm{e}^{\frac{\alpha_n}{R}}} |\widehat{\nabla g_n}(\xi)|^2\,d\xi\,.$$

Thanks to (40), $\int_{|\xi|\leq 1}|\widehat{\nabla g_n}(\xi)|^2\,d\xi \overset{n\to\infty}{\longrightarrow} 0$ and in view of (42)

$$\int_{1\leq|\xi|\leq \mathrm{e}^{\frac{\alpha_n}{R}}}|\widehat{\nabla g_n}(\xi)|^2\,d\xi \leq \frac{C_\psi}{\alpha_n}\int_1^{\mathrm{e}^{\frac{\alpha_n}{R}}}\frac{dr}{r}\leq\frac{C_\psi}{R}\overset{R\to\infty}{\longrightarrow}0\cdot$$

Secondly arguing as above, we write

$$|\xi|\,\widehat{g_n}(\xi)=I_1+I_2\,.$$

The fact that $\psi'\in\mathcal{D}(]0,+\infty[)$ implies that $I_2=0$, for $|\xi|\geq \mathrm{e}^{R\,\alpha_n}$ with R sufficiently large. Besides,

$$\begin{aligned}I_1&=\frac{1}{|\xi|}\sqrt{\frac{2\pi}{\alpha_n}}\int_0^{\frac{1}{\alpha_n}\log(|\xi|)}\partial_s\Big(J_0(\mathrm{e}^{-\alpha_n\,s}|\xi|)\Big)\,\psi'(s)\,ds\\&=\frac{1}{|\xi|}\sqrt{\frac{2\pi}{\alpha_n}}\left(\Big[J_0(\mathrm{e}^{-\alpha_n\,s}|\xi|)\,\psi'(s)\Big]_0^{\frac{1}{\alpha_n}\log(|\xi|)}-\int_0^{\frac{1}{\alpha_n}\log(|\xi|)}J_0(\mathrm{e}^{-\alpha_n\,s}|\xi|)\,\psi''(s)\,ds\right)\cdot\end{aligned}$$

Since $\psi'\in\mathcal{D}(]0,+\infty[)$ and $\frac{1}{\alpha_n}\log(|\xi|)\geq R$, we infer that for R large enough

$$\Big[J_0(\mathrm{e}^{-\alpha_n\,s}|\xi|)\,\psi'(s)\Big]_0^{\frac{1}{\alpha_n}\log(|\xi|)}=0\cdot$$

Finally if $[a,b]$ designates the support of ψ', we get in view of (33)

$$\begin{aligned}&\int_0^{\frac{1}{\alpha_n}\log(|\xi|)}J_0(\mathrm{e}^{-\alpha_n\,s}|\xi|)\,\psi''(s)\,ds\\&=\int_0^b\sqrt{\frac{2}{\pi\mathrm{e}^{-\alpha_n\,s}|\xi|}}\cos\left(\mathrm{e}^{-\alpha_n\,s}|\xi|-\frac{\pi}{4}\right)\psi''(s)\,ds+\int_0^b\mathrm{e}^{\frac{3}{2}\alpha_n\,s}O(|\xi|^{-\frac{3}{2}})\,\psi''(s)\,ds\cdot\end{aligned}$$

This gives rise to

$$\Big|\xi\,\widehat{g_n}(\xi)\Big|\leq\frac{C\|\psi''\|_{L^\infty}}{\alpha_n\sqrt{\alpha_n}\,|\xi|}\left(\frac{\mathrm{e}^{\frac{b}{2}\,\alpha_n}}{\sqrt{|\xi|}}+\frac{\mathrm{e}^{\frac{3\,b}{2}\alpha_n}}{|\xi|^{\frac{3}{2}}}\right)\cdot$$

We deduce that for $|\xi|\geq\mathrm{e}^{R\,\alpha_n}$

$$\int_{\frac{1}{\alpha_n}(\log|\xi|)\geq R}|\widehat{\nabla g_n}(\xi)|^2\,d\xi\leq\frac{C_\psi}{\alpha_n^3}\left(\mathrm{e}^{b\,\alpha_n}\int_{|\xi|\geq\mathrm{e}^{R\,\alpha_n}}\frac{d\xi}{|\xi|^3}+\mathrm{e}^{3\,b\,\alpha_n}\int_{|\xi|\geq\mathrm{e}^{R\,\alpha_n}}\frac{d\xi}{|\xi|^5}\right)\cdot$$

Thus for R large enough

$$\int_{\frac{1}{\alpha_n}(\log|\xi|)\geq R}|\widehat{\nabla g_n}(\xi)|^2\,d\xi\overset{n\to\infty}{\longrightarrow}0,$$

which achieves the proof of the proposition.

Acknowledgment

We are very grateful to P. Gérard for interesting discussions around the questions dealt with in this paper.

References

[1] S. Adachi and K. Tanaka, *Trudinger type inequalities in* $\mathbb{R}^N$ *and their best exponents*, Proceedings of the American Mathematical Society, **128 (7)** (2000), 2051–2057.

[2] Adimurthi and K. Tintarev, *On compactness in the Trudinger–Moser inequality*, arXiv:1110.3647, (2011).

[3] H. Bahouri, J.-Y. Chemin and R. Danchin, *Fourier Analysis and Nonlinear Partial Differential Equations*, Grundlehren der mathematischen Wissenschaften, Springer, (2011).

[4] H. Bahouri, A. Cohen and G. Koch, *A general wavelet-based profile decomposition in the critical embedding of function spaces*, Confluentes Mathematici, **3** (3) (2011), 1–25.

[5] H. Bahouri and I. Gallagher, *Weak stability of the set of global solutions to the Navier-Stokes equations*, arXiv:1109.4043, (2011).

[6] H. Bahouri and P. Gérard, *High frequency approximation of solutions to critical nonlinear wave equations*, American Journal of Math, **121** (1999), 131–175.

[7] H. Bahouri, M. Majdoub and N. Masmoudi, *On the lack of compactness in the* 2D *critical Sobolev embedding*, Journal of Functional Analysis, **260** (2011), 208–252.

[8] H. Bahouri, M. Majdoub and N. Masmoudi, *Lack of compactness in the 2D critical Sobolev embedding, the general case, Notes aux Comptes-Rendus de l'Académie des Sciences de Paris*, Ser. I **350**, 2012, pages 177–181.

[9] H. Bahouri, M. Majdoub and N. Masmoudi, *Lack of compactness in the 2D critical Sobolev embedding, the general case*, arXiv:1112.2998v1.

[10] J. Ben Ameur, Description du défaut de compacité de l'injection de Sobolev sur le groupe de Heisenberg, *Bulletin de la Société Mathématique de Belgique*, **15-4**, 2008, pages 599–624.

[11] H. Brezis and J.M. Coron, Convergence of solutions of H-Systems or how to blow bubbles, *Archiv for Rational Mechanics and Analysis*, **89**, 1985, pages 21–86.

[12] P. Gérard, *A microlocal version of concentration-compactness*, Partial differential equations and mathematical physics (Copenhagen, 1995; Lund, 1995), 143–157, Progr. Nonlinear Differential Equations Appl., **21**, Birkhäuser Boston, Boston, MA, (1996).

[13] P. Gérard, *Description du défaut de compacité de l'injection de Sobolev*, ESAIM: Control, Optimisation and Calculus of Variations, **3** (1998), 213–233 (electronic, URL: http://www.emath.fr/cocv/).

[14] A. Henrot and M. Pierre, *Variations et optimistation de formes*, Mathématiques et applications, Springer, **48**, (2005).

[15] S. Ibrahim and M. Majdoub, *Comparaison des ondes linéaires et non linéaires à coefficients variables*, Bulletin de la Société Mathématique de Belgique, **10** (2003), 299–312.

[16] S. Jaffard, *Analysis of the lack of compactness in the critical Sobolev embeddings*, Journal of Functional Analysis, **161** (1999), 384–396.

[17] C.E. Kenig and F. Merle, *Global well-posedness, scattering and blow-up for the energy critical focusing non-linear wave equation*, Acta Mathematica, **201** (2008), 147–212.

[18] S. Keraani, *On the defect of compactness for the Strichartz estimates of the Schrödinger equation*, Journal of Differential equations, **175** (2001), 353–392.

[19] C. Laurent, *On stabilization and control for the critical Klein–Gordon equation on a 3D compact manifold*, to appear in Journal of Functional Analysis.

[20] N.N. Lebedev, *Special functions and their applications*, Prentice Hall, 1965.

[21] P.-L. Lions, *The concentration-compactness principle in the calculus of variations. The limit case. I.*, Revista Matemática Iberoamericana, **1** (1985), 145–201.

[22] P.-L. Lions, *The concentration-compactness principle in the calculus of variations. The locally compact case, I*, Ann. Inst. H. Poincaré Anal. Non Linéaire 1 (1984), 109–145.

[23] G. Mancini and K. Sandeep, *Moser–Trudinger inequality on conformal discs*, Communications in Contemporary Mathematics **12** **(6)** (2010), 1055–1068.

[24] J. Moser, *A sharp form of an inequality of N. Trudinger*, Indiana University Mathematics Journal, **20** (1971), 1077–1092.

[25] F.W.J. Olver, *Asymptotics and special functions*, Academic Press, 1974.

[26] M.-M. Rao and Z.-D. Ren, *Applications of Orlicz spaces*, Monographs and Textbooks in Pure and Applied Mathematics, **250** (2002), Marcel Dekker Inc.

[27] B. Ruf, *A sharp Trudinger–Moser type inequality for unbounded domains in* $\mathbb{R}^2$, Journal of Functional Analysis, **219** (2005), 340–367.

[28] I. Schindler and K. Tintarev, An abstract version of the concentration compactness principle, *Revista Math. Complutense*, **15-2**, 2002, pages 417–436.

[29] S. Solimini, *A note on compactness-type properties with respect to Lorentz norms of bounded subset of a Sobolev space*, Annales de l'IHP analyse non linéaire, **12** (1995), 319–337.

[30] M. Struwe, *A global compactness result for boundary value problems involving limiting nonlinearities*, Mathematische Zeitschrift, 187, 511–517, 1984.

Hajer Bahouri
LAMA, Université Paris-Est Créteil
61, avenue du Général de Gaulle
F-94010 Créteil Cedex, France
e-mail: `hbahouri@math.cnrs.fr`

Concentration Analysis and Applications to PDE
Trends in Mathematics, 17–28

A Class of Second-order Dilation Invariant Inequalities

Paolo Caldiroli and Roberta Musina

Abstract. We compute the best constants in some dilation invariant inequalities for the weighted L^2-norms of $-\Delta u$ and ∇u, with weights being powers of the distance from the origin.

Mathematics Subject Classification (2010). 26D10, 47F05.

Keywords. Rellich inequality, dilation invariant inequalities, critical dimensions, weighted biharmonic operator.

1. Introduction

In recent years, there has been a growing interest in dilation invariant inequalities that are somehow related with the famous Rellich inequality [15], [16]. We shall not attempt to provide a complete list of references on this subject. However, among the more recent contributions we cite [1]–[13], [17] and references therein.

In the present paper we study a class of inequalities for the weighted L^2-norms of $-\Delta u$ and ∇u. More precisely, let $n \ge 2$ be a given integer, let $\alpha \in \mathbb{R}$ be a varying parameter, and let Σ be a regular domain in $\mathbb{S}^{n-1}$. We are interested in inequalities of the form

$$\int_{\mathcal{C}_\Sigma} |x|^\alpha |\Delta u|^2 dx \ge c \int_{\mathcal{C}_\Sigma} |x|^{\alpha-2} |\nabla u|^2 dx \quad \text{for any } u \in C^2_c(\overline{\mathcal{C}_\Sigma} \setminus \{0\}) \tag{1.1}$$

where $\mathcal{C}_\Sigma$ denotes the cone in $\mathbb{R}^n$ spanned by Σ, namely

$$\mathcal{C}_\Sigma = \left\{ x \in \mathbb{R}^n \setminus \{0\} \;\middle|\; \frac{x}{|x|} \in \Sigma \right\}.$$

The first author is partially supported by the PRIN2009 grant "Critical Point Theory and Perturbative Methods for Nonlinear Differential Equations". The second author is partially supported by Miur-PRIN 2009WRJ3W7-001 "Fenomeni di concentrazione e problemi di analisi geometrica".

Notice that $\mathcal{C}_\Sigma = \mathbb{R}^n \setminus \{0\}$ when $\Sigma = \mathbb{S}^{n-1}$. Our aim is to compute the best constant

$$\delta_{n,\alpha}(\mathcal{C}_\Sigma) := \inf_{\substack{u \in C^2_c(\overline{\mathcal{C}_\Sigma} \setminus \{0\}) \\ u \neq 0}} \frac{\displaystyle\int_{\mathcal{C}_\Sigma} |x|^\alpha |\Delta u|^2 dx}{\displaystyle\int_{\mathcal{C}_\Sigma} |x|^{\alpha-2} |\nabla u|^2 dx} .$$

In fact this goal was already accomplished by Ghoussoub and Moradifam in [11] in the case of the whole space. However we provide alternative proofs which may be naturally adapted to handle cone like domains. Even if this generalization to cones seems to have a somehow artificial flavour, in fact in our opinion it contains some deeper features. Firstly it allows us to consider the case of domains involving a singularity on the boundary, like the half-space. Moreover our results are stated in a fashion which makes clearer the expression of the best constant even in the case of the whole space. This fact is strongly related to the particular approach followed here.

We also mention the papers [2] and [14] dealing with a class of inequalities for radially symmetric functions on $\mathbb{R}^n$ in the non-Hilbertian case, that is, involving the weighted L^p-norms of $-\Delta u$ and ∇u, with $p > 1$.

In order to state our main results we put

$$\gamma_{n,\alpha} = \frac{(n-4+\alpha)(n-\alpha)}{4} , \quad h_{n,\alpha} = \left(\frac{n-4+\alpha}{2} \right)^2 .$$

Given a domain Σ in $\mathbb{S}^{n-1}$ with $\partial\Sigma \in C^2$, we denote by Λ_Σ the spectrum of the Laplace–Beltrami operator on Σ with null boundary conditions and by λ_Σ the first eigenvalue. Notice that $\lambda_\Sigma > 0$ apart from the case $\Sigma = \mathbb{S}^{n-1}$.

Theorem 1.1. *Let $n \geq 2$ and let Σ be a domain in $\mathbb{S}^{n-1}$ with $\partial\Sigma \in C^2$. Assume $\alpha \neq 4-n$. Then the following facts hold.*

(i) $\delta_{n,\alpha}(\mathcal{C}_\Sigma) > 0$ *if and only if* $-\gamma_{n,\alpha} \notin \Lambda_\Sigma$. *Moreover*

$$\delta_{n,\alpha}(\mathcal{C}_\Sigma) \leq M_{n,\alpha}(\Sigma) := \min_{\lambda \in \Lambda_\Sigma} \frac{(\gamma_{n,\alpha} + \lambda)^2}{h_{n,\alpha} + \lambda} . \tag{1.2}$$

(ii) *If* $\gamma_{n,\alpha} - 2h_{n,\alpha} \leq \lambda_\Sigma$ *then* $\delta_{n,\alpha}(\mathcal{C}_\Sigma) = M_{n,\alpha}(\Sigma)$.

When $\Sigma = \mathbb{S}^{n-1}$ we can be more precise. First of all, as well as in [11], we have the following sharp result for the best constant in the class of radially symmetric functions.

Theorem 1.2. *Let $n \geq 2$ and $\alpha \in \mathbb{R}$. Then*

$$\delta^{\rm rad}_{n,\alpha} := \inf_{\substack{u \in C^2_c(\mathbb{R}^n \setminus \{0\}) \\ u = u(|x|) \ , \ u \neq 0}} \frac{\displaystyle\int_{\mathbb{R}^n} |x|^\alpha |\Delta u|^2 dx}{\displaystyle\int_{\mathbb{R}^n} |x|^{\alpha-2} |\nabla u|^2 dx} = \left(\frac{n-\alpha}{2} \right)^2 .$$

Theorem 1.2 will be proved in Section 2.3. When we allow u to be any function in $C_c^2(\mathbb{R}^n \setminus \{0\})$ non necessarily radial we can estimate the best constant with the aid of Theorem 1.1 and using the explicit knowledge of the spectrum of the Laplace–Beltrami operator on the sphere:

$$\Lambda_{\mathbb{S}^{n-1}} = \{k(n-2+k) \mid k \in \mathbb{N} \cup \{0\}\}.$$

In particular $\lambda_{\mathbb{S}^{n-1}} = 0$. To simplify the notation, we write $\delta_{n,\alpha}$ instead of $\delta_{n,\alpha}(\mathbb{R}^n \setminus \{0\})$, and $M_{n,\alpha}$ instead of $M_{n,\alpha}(\mathbb{S}^{n-1})$. The results stated in the next theorems are already known (see [11]) but we prove them in a different way.

Theorem 1.3. *Let $n \geq 2$ and assume $\alpha \neq 4-n$.*

(i) *If $n = 2$ then $\delta_{2,\alpha} = M_{2,\alpha}$ for any $\alpha \in \mathbb{R}$.*
(ii) *If $n \geq 3$ then there exists $\alpha^* \in [4-n, 2)$ such that $\delta_{n,\alpha} = M_{n,\alpha}$ for any $\alpha \notin [4-n, \alpha^*)$.*
(iii) *If $n \geq 3$ and $\alpha^* < \alpha < n$ then $\delta_{n,\alpha} = \delta_{n,\alpha}^{\mathrm{rad}}$.*

In the "*critical case*" $\alpha = 4-n$ a very singular phenomenon can be observed.

Theorem 1.4. *If $\alpha = 4-n$ then*

$$\delta_{n,4-n} = \min\left\{(n-2)^2\ ,\ n-1\right\}.$$

In particular,

$$\delta_{n,4-n} > 0 \text{ for any } n \geq 3 \text{ and } \delta_{n,4-n} = n-1 < \delta_{n,4-n}^{\mathrm{rad}} \text{ for any } n \geq 4.$$

It should be emphasized that, apart from the two-dimensional case, the function $\alpha \mapsto \delta_{n,\alpha}$ is not continuous at $\alpha = 4-n$, since $\gamma_{n,4-n} = h_{n,4-n} = 0$ and (1.2) implies that $\delta_{n,\alpha} \to 0$ as $\alpha \to 4-n$.

Let us make some remarks about the above results in the meaningful case $\alpha = 0$. First notice that in two dimensions $\delta_{2,0} = 0 < \delta_{2,0}^{\mathrm{rad}} = 1$. In dimension $n = 3$ the best constant $\delta_{3,0}$, already known according to the paper [11] can be computed by means of the formula for $M_{3,0}$ and yields:

$$\int_{\mathbb{R}^3} |\Delta u|^2\,dx \geq \frac{25}{36}\int_{\mathbb{R}^3} |x|^{-2}|\nabla u|^2\,dx \quad \text{for any } u \in C_c^2(\mathbb{R}^3 \setminus \{0\}).$$

Notice that $\delta_{3,0}^{\mathrm{rad}} = 9/4$ is larger than the best constant on the whole space and breaking symmetry occurs. A similar phenomenon appears in the *critical dimension* $n = 4$. Indeed $\delta_{4,0}^{\mathrm{rad}} = 4$, while from Theorem 1.4 it follows that 3 is the best constant in the inequality

$$\int_{\mathbb{R}^4} |\Delta u|^2\,dx \geq 3\int_{\mathbb{R}^4} |x|^{-2}|\nabla u|^2\,dx \quad \text{for any } u \in C_c^2(\mathbb{R}^4 \setminus \{0\}).$$

To handle higher dimensions we estimate

$$\alpha^* < \frac{1}{3}\left(n + 4 - 2\sqrt{n^2 - n + 1}\right). \tag{1.3}$$

Notice that $\alpha^* < 0$ if $n \geq 5$. A standard density result can be used to infer the next corollary.

Corollary 1.5. *Assume $n \ge 5$. Then*

$$\int_{\mathbb{R}^n} |\Delta u|^2 \, dx \ge \frac{n^2}{4} \int_{\mathbb{R}^n} |x|^{-2} |\nabla u|^2 \, dx \quad \text{for any } u \in \mathcal{D}^{2,2}(\mathbb{R}^n),$$

and $n^2/4$ is the best constant.

2. Proofs

The only tools we use are the Cauchy–Schwarz inequality, integration by parts, the variational characterization of the eigenvalues, and the Emden–Fowler transform $u \mapsto w = Tu$, that is defined via

$$u(x) = |x|^{\frac{4-n-\alpha}{2}} \, w\left(-\log |x|, \frac{x}{|x|} \right).$$

Such a transform T maps functions $u \colon \mathbb{R}^n \setminus \{0\} \to \mathbb{R}$ into functions $w = w(s, \sigma)$ on the cylinder $\mathbb{R} \times \mathbb{S}^{n-1}$. More generally, given a domain Σ in $\mathbb{S}^{n-1}$, let us denote

$$\mathcal{Z}_\Sigma := \mathbb{R} \times \Sigma$$

the corresponding cylinder. We point out that $w \in C^2_c(\overline{\mathcal{Z}_\Sigma})$ as $u \in C^2_c(\overline{\mathcal{C}_\Sigma} \setminus \{0\})$. Moreover, by direct computation (see for instance [6]), it can be proved that

$$\delta_{n,\alpha}(\mathcal{C}_\Sigma) = \inf_{\substack{w \in C^2_c(\overline{\mathcal{Z}_\Sigma}) \\ w \ne 0}} \frac{\displaystyle\int_{\mathcal{Z}_\Sigma} \left| \Delta_\sigma w + w_{ss} + (\alpha - 2) w_s - \gamma_{n,\alpha} w \right|^2 ds d\sigma}{\displaystyle\int_{\mathcal{Z}_\Sigma} \left(|\nabla_\sigma w|^2 + |w_s|^2 \right) ds d\sigma + h_{n,\alpha} \int_{\mathcal{Z}_\Sigma} |w|^2 ds d\sigma}.$$

Here and in the rest of the paper we denote by $-\Delta_\sigma$, ∇_σ the Laplace–Beltrami operator and the gradient on $\mathbb{S}^{n-1}$, respectively, while w_s is the derivative of w with respect to $s \in \mathbb{R}$.

2.1. Some notation and technical lemmas

For every eigenvalue $\lambda \in \Lambda_\Sigma$ let

$$Y_\lambda := \{ g\varphi \mid g \in C^2_c(\mathbb{R}), \ \varphi \text{ eigenfunction corresponding to } \lambda \}.$$

Notice that $Y_\lambda \subset C^2_c(\overline{\mathcal{Z}_\Sigma})$. Moreover set

$$V_\lambda := \{ v \in C^2_c(\overline{\mathcal{Z}_\Sigma}) \mid \int_{\mathcal{Z}_\Sigma} vw \, ds d\sigma = 0 \ \forall w \in Y_{\lambda'}, \forall \lambda' \in \Lambda_\Sigma, \ \lambda' < \lambda \}.$$

Hence $V_\lambda \supset Y_\lambda$ and $V_\lambda = C^2_c(\overline{\mathcal{Z}_\Sigma})$ when $\lambda = \lambda_\Sigma$.

Lemma 2.1. *For every $\lambda \in \Lambda_\Sigma$ one has that*

$$\inf_{\substack{w \in V_\lambda \\ w \ne 0}} \frac{\displaystyle\int_{\mathcal{Z}_\Sigma} \left(|\nabla_\sigma w|^2 + |w_s|^2 \right) ds d\sigma}{\displaystyle\int_{\mathcal{Z}_\Sigma} |w|^2 \, ds d\sigma} \ge \lambda.$$

Proof. Clearly

$$\inf_{\substack{w\in V_\lambda\\ w\neq 0}} \frac{\int_{\mathcal{Z}_\Sigma} \left(|\nabla_\sigma w|^2 + |w_s|^2\right) dsd\sigma}{\int_{\mathcal{Z}_\Sigma} |w|^2\, dsd\sigma} = \inf_{\substack{w\in V_\lambda\\ w\neq 0}} \frac{\int_{\mathcal{Z}_\Sigma} |\nabla_\sigma w|^2\, dsd\sigma}{\int_{\mathcal{Z}_\Sigma} |w|^2\, dsd\sigma}.$$

Then the conclusion follows from the fact that every mapping $w \in V_\lambda$ is orthogonal to $Y_{\lambda'}$ for any eigenvalue $\lambda' < \lambda$ and from the variational characterization of the eigenvalues. □

For $A, B, C \in \mathbb{R}$ set

$$\begin{aligned} N_{A,B}(w) &= \int_{\mathcal{Z}_\Sigma} |\Delta_\sigma w + w_{ss} + Aw_s - Bw|^2\, dsd\sigma \\ D_C(w) &= \int_{\mathcal{Z}_\Sigma} \left(|\nabla_\sigma w|^2 + |w_s|^2\right) dsd\sigma + C\int_{\mathcal{Z}_\Sigma} |w|^2\, dsd\sigma . \end{aligned} \tag{2.1}$$

Moreover, for $\lambda \in \Lambda_\Sigma$, set

$$M_\lambda(A,B,C) = \inf_{\substack{w\in V_\lambda\\ w\neq 0}} \frac{N_{A,B}(w)}{D_C(w)} \quad \text{and} \quad \widetilde{M}_\lambda(A,B,C) = \inf_{\substack{w\in Y_\lambda\\ w\neq 0}} \frac{N_{A,B}(w)}{D_C(w)} .$$

Lemma 2.2. *For every $\lambda \in \Lambda_\Sigma$, if $0 < B + \lambda \le 2(C + \lambda)$, then*

$$M_\lambda(A,B,C) \ge \frac{(B+\lambda)^2}{C+\lambda} .$$

Proof. For every $w \in V_\lambda$, integrating by parts and using the Cauchy–Schwarz inequality, we obtain

$$\begin{aligned} D_B(w) &= -\int_{\mathcal{Z}_\Sigma} w\left(\Delta_\sigma w + w_{ss} + Aw_s - Bw\right) dsd\sigma \\ &\le \left(\int_{\mathcal{Z}_\Sigma} |w|^2\, dsd\sigma\right)^{\frac{1}{2}} \left(N_{A,B}(w)\right)^{\frac{1}{2}} . \end{aligned}$$

Then for $w \in V_\lambda \setminus \{0\}$ we have

$$\frac{N_{A,B}(w)}{D_C(w)} \ge \frac{(R(w)+B)^2}{R(w)+C} \quad \text{where} \quad R(w) = \frac{\int_{\mathcal{Z}_\Sigma} \left(|\nabla_\sigma w|^2 + |w_s|^2\right) dsd\sigma}{\int_{\mathcal{Z}_\Sigma} |w|^2\, dsd\sigma}.$$

Therefore, using Lemma 2.1, we infer that

$$M_\lambda(A,B,C) \ge \inf_{r\ge\lambda} \frac{(B+r)^2}{C+r} = \frac{(B+\lambda)^2}{C+\lambda}$$

where the last equality can be obtained by elementary calculus using the assumptions on B and C. □

Lemma 2.3. *For every $\lambda \in \Lambda_\Sigma$, if $A^2+2(B+\lambda) > (B+\lambda)^2/(C+\lambda)$ and $C+\lambda > 0$, then*

$$\widetilde{M}_\lambda(A,B,C) \geq \frac{(B+\lambda)^2}{C+\lambda}\,.$$

Proof. Using the definition of Y_λ we obtain that

$$\widetilde{M}_\lambda(A,B,C) = \inf_{\substack{g\in C^2_c(\mathbb{R})\\ g\neq 0}} \frac{\displaystyle\int_{-\infty}^{\infty} |g''+Ag'-(B+\lambda)g|^2\,ds}{\displaystyle\int_{-\infty}^{\infty} \left(|g'|^2+(C+\lambda)|g|^2\right)\,ds}\,.$$

To simplify notation, we can assume that

$$\int_{-\infty}^{\infty} |g|^2\,ds = 1.$$

Integration by parts yields

$$\begin{aligned}&\int_{-\infty}^{\infty} |g''+Ag'-(B+\lambda)g|^2 ds\\ &\quad= \int_{-\infty}^{\infty} |g''|^2\,ds + (A^2+2(B+\lambda))\int_{-\infty}^{\infty} |g'|^2\,ds + (B+\lambda)^2.\end{aligned}$$

Moreover by the Cauchy–Schwarz and the Young inequality we estimate

$$\int_{-\infty}^{\infty} |g'|^2\,ds = -\int_{-\infty}^{\infty} g''g\,ds \leq \left(\int_{-\infty}^{\infty} |g''|^2\,ds\right)^{\frac12} \leq \frac{\varepsilon}{2} + \frac{1}{2\varepsilon}\int_{-\infty}^{\infty} |g''|^2\,ds\,.$$

Then

$$\begin{aligned}&\frac{\displaystyle\int_{-\infty}^{\infty} |g''+Ag'-(B+\lambda)g|^2\,ds}{\displaystyle\int_{-\infty}^{\infty} \left(|g'|^2+(C+\lambda)|g|^2\right)\,ds}\\ &\qquad\geq \frac{(A^2+2(B+\lambda+\varepsilon))\displaystyle\int_{-\infty}^{\infty} |g'|^2\,ds + (B+\lambda)^2-\varepsilon^2}{\displaystyle\int_{-\infty}^{\infty} |g'|^2\,ds + C+\lambda}\end{aligned}$$

and consequently

$$\widetilde{M}_\lambda(A,B,C) \geq (A^2+2(B+\lambda+\varepsilon))\inf_{t\geq 0}\frac{B_\varepsilon+t}{C+\lambda+t}$$

where

$$B_\varepsilon = \frac{(B+\lambda)^2-\varepsilon^2}{A^2+2(B+\lambda+\varepsilon)}.$$

By the assumptions on A, B and C we have that $C+\lambda > B_\varepsilon > 0$ for $\varepsilon > 0$ small enough. Then, by elementary calculus,

$$\inf_{t\geq 0}\frac{B_\varepsilon+t}{C+\lambda+t} = \frac{B_\varepsilon}{C+\lambda}.$$

Hence for $\varepsilon > 0$ small enough

$$\widetilde{M}_\lambda(A,B,C) \geq \frac{(B+\lambda)^2 - \varepsilon^2}{C+\lambda}$$

and letting $\varepsilon \to 0$ we get the conclusion. □

2.2. Proof of Theorem 1.1

Fix $\alpha \in \mathbb{R}$ and $n \in \mathbb{N}$, $n \geq 2$. For every $w \in C^2_c(\overline{\mathcal{Z}_\Sigma})$ set

$$N(w) = \int_{\mathcal{Z}_\Sigma} |\Delta_\sigma w + w_{ss} + (\alpha-2)w_s - \gamma_{n,\alpha} w|^2 \, ds d\sigma$$

$$D(w) = \int_{\mathcal{Z}_\Sigma} \left(|\nabla_\sigma w|^2 + |w_s|^2\right) ds d\sigma + h_{n,\alpha} \int_{\mathcal{Z}_\Sigma} |w|^2 \, ds d\sigma \, .$$

Notice that according to the notation (2.1) we have that $N = N_{A,B}$ and $D = D_C$ with

$$A = \alpha - 2 \, , \quad B = \gamma_{n,\alpha} \, , \quad C = h_{n,\alpha} \, . \tag{2.2}$$

Proof of (i). Since $\alpha \neq 4-n$, then $h_{n,\alpha} > 0$ and therefore the functional D is the square of an equivalent Hilbertian norm on $H^1(\mathcal{Z}_\Sigma)$. Assume that $-\gamma_{n,\alpha} \notin \Lambda_\Sigma$. In this case, by the results in [6], the functional N is the square of an equivalent Hilbertian norm on $H^2(\mathcal{Z}_\Sigma)$. Therefore, since with the above notation

$$\delta_{n,\alpha}(\mathcal{C}_\Sigma) = \inf\left\{\frac{N(w)}{D(w)} \mid w \in C^2_c(\overline{\mathcal{Z}_\Sigma}), \ w \neq 0\right\},$$

as $H^2(\mathcal{Z}_\Sigma)$ is continuously embedded into $H^1(\mathcal{Z}_\Sigma)$, we obtain $\delta_{n,\alpha}(\mathcal{C}_\Sigma) > 0$. The fact that $\delta_{n,\alpha}(\mathcal{C}_\Sigma) = 0$ if $-\gamma_{n,\alpha} \in \Lambda_\Sigma$ is a consequence of (1.2). To check (1.2) we fix $\lambda \in \Lambda_\Sigma$ and we estimate

$$\begin{aligned}
\delta_{n,\alpha}(\mathcal{C}_\Sigma) &\leq \inf_{\substack{w \in Y_\lambda \\ w \neq 0}} \frac{N(w)}{D(w)} \\
&= \inf_{\substack{g \in C^2_c(\mathbb{R}) \\ g \neq 0}} \frac{\displaystyle\int_{\mathbb{R}} |g'' + (\alpha-2)g' - (\gamma_{n,\alpha}+\lambda)g|^2 \, ds}{\displaystyle\int_{\mathbb{R}} |g'|^2 ds + (h_{n,\alpha}+\lambda)\int_{\mathbb{R}} |g|^2 ds} = \frac{(\gamma_{n,\alpha}+\lambda)^2}{h_{n,\alpha}+\lambda}.
\end{aligned}$$

The last equality can be easily checked taking $g(s) = g_0(\varepsilon s)$ with $g_0 \in C^2_c(\mathbb{R})$ fixed, $g_0 \neq 0$, and $\varepsilon > 0$, and letting $\varepsilon \to 0$. Then (1.2) follows from the arbitrariness of $\lambda \in \Lambda_\Sigma$.

Proof of (ii). It suffices to study the case $-\gamma_{n,\alpha} \notin \Lambda_\Sigma$, since otherwise $\delta_{n,\alpha}(\mathcal{C}_\Sigma) = M_{n,\alpha}(\Sigma) = 0$. Let us distinguish the argument according that $-\gamma_{n,\alpha}$ stays below the spectrum or not.

Case $-\gamma_{n,\alpha} < \lambda_\Sigma$. Since $C^2_c(\overline{\mathcal{Z}_\Sigma}) = V_{\lambda_\Sigma}$, we have that $\delta_{n,\alpha}(\mathcal{C}_\Sigma) = M_{\lambda_\Sigma}(A,B,C)$ with A, B, and C given as in (2.2). We apply Lemma 2.2 with $\lambda = \lambda_\Sigma$. The condition $B + \lambda > 0$ is fulfilled since we are dealing with the case $-\gamma_{n,\alpha} < \lambda_\Sigma$. The condition $B + \lambda \leq 2(C+\lambda)$ is equivalent to say $\gamma_{n,\alpha} \leq 2h_{n,\alpha} + \lambda_\Sigma$. Hence if

$-\lambda_\Sigma < \gamma_{n,\alpha} \le 2h_{n,\alpha} + \lambda_\Sigma$ then

$$\delta_{n,\alpha}(\mathcal{C}_\Sigma) \ge \frac{(\gamma_{n,\alpha} + \lambda_\Sigma)^2}{h_{n,\alpha} + \lambda_\Sigma} \ge M_{n,\alpha}(\Sigma).$$

Hence, in this case, by (1), $\delta_{n,\alpha}(\mathcal{C}_\Sigma) = M_{n,\alpha}(\Sigma)$.

Case $-\gamma_{n,\alpha} > \lambda_\Sigma$. We can find two consecutive eigenvalues λ_{k-1} and λ_k such that

$$\lambda_{k-1} < -\gamma_{n,\alpha} < \lambda_k\,.$$

Any $w \in C^2_c(\overline{\mathcal{Z}_\Sigma})$ can be written according to the following decomposition

$$w = v_1 + \cdots + v_k$$

with $v_j \in Y_{\lambda_j}$ for $j = 1, \dots, k-1$, and $v_k \in V_{\lambda_k}$. One easily checks that

$$\frac{N(w)}{D(w)} = \sum_{j=1}^k \theta_j \frac{N(v_j)}{D(v_j)} \quad \text{where} \quad \theta_j = \frac{D(v_j)}{D(w)}.$$

Since $\theta_j \ge 0$ for all $j = 1, \dots, k$ and $\theta_1 + \cdots + \theta_k = 1$, we have that

$$\frac{N(w)}{D(w)} \ge \min_{j=1,\dots,k} \frac{N(v_j)}{D(v_j)}. \tag{2.3}$$

We estimate $N(v_j)/D(v_j)$ for $j = 1, \dots, k-1$ by means of Lemma 2.3 with $\lambda = \lambda_j$ and A, B, and C as in (2.2). The condition $C + \lambda > 0$ is fulfilled as $\lambda_j \ge 0$ and $h_{n,\alpha} > 0$ since, by hypothesis, $\alpha \ne 4 - n$. The condition $A^2 + 2(B + \lambda) > (B + \lambda)^2/(C + \lambda)$ can be checked by considering the function

$$\Phi(t) = \left(2t + \frac{(n-2)^2}{2} + \frac{(\alpha-2)^2}{2}\right)(t + h_{n,\alpha}) - (t + \gamma_{n,\alpha})^2.$$

One has that $\Phi(0) = h^2_{n,\alpha} > 0$ and $\Phi'(0) = 2h_{n,\alpha} + (\alpha - 2)^2 > 0$. Then $\Phi(t) > 0$ for all $t \ge 0$. In particular $\Phi(\lambda_j) > 0$ and

$$A^2 + 2(B + \lambda_j) - \frac{(B + \lambda_j)^2}{C + \lambda_j} = \frac{\Phi(\lambda_j)}{h_{n,\alpha} + \lambda_j} > 0.$$

Hence Lemma 2.3 applies and yields

$$\frac{N(v_j)}{D(v_j)} \ge \frac{(\gamma_{n,\alpha} + \lambda_j)^2}{h_{n,\alpha} + \lambda_j} \ge M_{n,\alpha}(\Sigma) \quad \forall j = 1, \dots, k-1. \tag{2.4}$$

In order to estimate $N(v_k)/D(v_k)$ we apply Lemma 2.2 with $\lambda = \lambda_k$ and A, B, and C as in (2.2). The condition $B + \lambda > 0$ is satisfied since $-\gamma_{N,\alpha} < \lambda_k$. The other condition $B + \lambda \le 2(C + \lambda)$ is also fulfilled since

$$2(C + \lambda) - (B + \lambda) = 2h_{n,\alpha} - \gamma_{n,\alpha} + \lambda_k > 2h_{n,\alpha} - \gamma_{n,\alpha} + \lambda_\Sigma > 0$$

by the assumption made in **(ii)**. Therefore Lemma 2.2 applies and thus

$$\frac{N(v_k)}{D(v_k)} \ge \frac{(\gamma_{n,\alpha} + \lambda_k)^2}{h_{n,\alpha} + \lambda_k} \ge M_{n,\alpha}. \tag{2.5}$$

In conclusion by (2.3)–(2.5) and by the arbitrariness of $w \in C_c^2(\overline{\mathcal{Z}_\Sigma})$ one concludes as in the first case. □

2.3. Proof of Theorem 1.2

For a fixed radial function $u \in C_c^2(\mathbb{R}^n \setminus \{0\})$ we introduce the radially symmetric function

$$v(x) = |x|^{\frac{n+\alpha-2}{2}} u_r(x),$$

where u_r is the radial derivative of u. Then

$$\begin{aligned}\int_{\mathbb{R}^n} |x|^\alpha |\Delta u|^2\, dx &= \int_{\mathbb{R}^n} |x|^\alpha \left| \frac{n-\alpha}{2} |x|^{-1} u_r + |x|^{\frac{2-n-\alpha}{2}} v_r \right|^2 dx \\ &= \left(\frac{n-\alpha}{2}\right)^2 \int_{\mathbb{R}^n} |x|^{\alpha-2} |\nabla u|^2\, dx + \int_{\mathbb{R}^n} |x|^{2-n} |\nabla v|^2\, dx,\end{aligned}$$

since the double product vanishes:

$$\int_{\mathbb{R}^n} |x|^{\frac{\alpha-n}{2}} v_r u_r\, dx = \int_{\mathbb{R}^n} |x|^{1-n} v v_r\, dx = c \int_0^\infty (v^2)_r\, dr = 0.$$

The conclusion is immediate. □

2.4. Proof of Theorem 1.3

We apply Theorem 1.1 considering that we deal with the case $\lambda_\Sigma = 0$.

Proof of (i). If $n = 2$ and $\alpha \neq 2$ then $\gamma_{n,\alpha} = -(\alpha-2)^2/4 < 0$. The condition

$$\gamma_{n,\alpha} - 2h_{n,\alpha} \le \lambda_\Sigma \tag{2.6}$$

holds true for every $\alpha \neq 2$ and thus one can conclude.

Proof of (ii). Consider now the case $n \ge 3$. Suppose $\gamma_{n,\alpha} > 0$, i.e., $\alpha \in (4-n, n)$. In this case the condition (2.6) holds true if and only if $\alpha \ge (n-8)/3$. When $\gamma_{n,\alpha} < 0$, the condition (2.6) always holds true. Hence (ii) is proved with $\alpha^* < (n-8)/3$.

Proof of (iii). If $n \ge 3$ and $\alpha \in (\alpha^*, n)$ then $\gamma_{n,\alpha} > 0$ and the mapping $t \mapsto (\gamma_{n,\alpha} + t)^2/(h_{n,\alpha} + t)$ is increasing in $[0, \infty)$. Hence $M_{n,\alpha} = \gamma_{n,\alpha}^2/h_{n,\alpha} = \delta_{n,\alpha}^{\mathrm{rad}}$, by Theorem 1.2. □

2.5. Proof of Theorem 1.4

First notice that $\delta_{n,4-n} \le \delta_{n,4-n}^{\mathrm{rad}} = (n-2)^2$ by Theorem 1.2. Now we prove that $\delta_{n,4-n} \le n-1$. Notice that $\gamma_{n,4-n} = h_{n,4-n} = 0$. We estimate $\delta_{n,4-n}$ with a family of mappings $w(s,\sigma) = g(\varepsilon s)\varphi(\sigma)$ where $g \in C_c^2(\mathbb{R})$ is any nontrivial fixed function, $\varepsilon > 0$ and φ is an eigenfunction for $-\Delta_\sigma$ on $\mathbb{S}^{n-1}$ relative to the first positive eigenvalue $(n-1)$. In this way we obtain

$$\delta_{n,4-n} \le \frac{\displaystyle\int_{\mathbb{R}} \left|\varepsilon^2 g'' + (\alpha-2)\varepsilon g' - (n-1) g\right|^2 ds}{\displaystyle\varepsilon^2 \int_{\mathbb{R}} |g'|^2\, ds + (n-1)\int_{\mathbb{R}} |g|^2\, ds}.$$

Then, passing to the limit as $\varepsilon \to 0$, we conclude that $\delta_{n,4-n} \le n-1$. Thus $\delta_{n,4-n} \le \min\{(n-2)^2 \ , \ n-1\}$. To prove the opposite inequality we argue by contradiction. We assume that there exists $w \in C^2_c(\mathbb{R} \times \mathbb{S}^{n-1})$, $w \neq 0$, such that

$$\min\{(n-2)^2 \ , \ n-1\} > \frac{N(w)}{D(w)}, \tag{2.7}$$

where $N(w)$ and $D(w)$ are as in the proof of Theorem 1.1. We can write w as $w(s,\sigma) = g(s) + v(s,\sigma)$, where

$$\int_{\mathbb{S}^{n-1}} v(s,\sigma) d\sigma = 0$$

for any $s \in \mathbb{R}$. Notice that $v \neq 0$, otherwise (2.7) would contradict Theorem 1.2. Thus

$$\xi_v := \frac{\displaystyle\int_{\mathcal{Z}} |\nabla_\sigma v|^2 \, ds d\sigma}{\displaystyle\int_{\mathcal{Z}} |v|^2 \, ds d\sigma} \ge n-1. \tag{2.8}$$

Clearly, $N(g) \ge (n-2)^2 D(g)$ by Theorem 1.2. Arguing as in the proof of Theorem 1.1 and using (2.8) we can estimate $N(v) \ge (n-1)D(v)$. Therefore

$$\begin{aligned} \min\{(n-2)^2 \ , \ n-1\} &> \frac{N(w)}{D(w)} \\ &= \frac{N(g) + N(v)}{D(g) + D(v)} \\ &\ge \frac{(n-2)^2 D(g) + (n-1) D(v)}{D(g) + D(v)}, \end{aligned}$$

that readily leads to a contradiction. Thus, the equality holds and the theorem is completely proved. □

2.6. Proof of (1.3)

Let $\alpha \in (4-n, 2)$ such that $\delta^{\mathrm{rad}}_{n,\alpha} > \delta_{n,\alpha}$. Thus there exist $g \in C^2_c(\mathbb{R})$ and $v \in C^2_c(\mathcal{Z})$ such that $v(s,\cdot)$ has zero mean value on the sphere for any $s \in \mathbb{R}$, and such that

$$\delta^{\mathrm{rad}}_{n,\alpha} > \frac{N(g+v)}{D(g+v)},$$

where $N(\cdot)$ and $D(\cdot)$ are defined as in the proof of Theorem 1.1. In addition, it holds that $v \neq 0$ and that v satisfies (2.8). Clearly, $N(g) \ge \delta^{\mathrm{rad}}_{n,\alpha} D(g)$. Arguing as in the proof of Theorem 1.1 we can estimate

$$\frac{N(v)}{D(v)} \ge \frac{(\xi_v + \gamma_{n,\alpha})^2}{\xi_v + h_{n,\alpha}},$$

where ξ_v is defined in (2.8). Therefore

$$\delta_{n,\alpha}^{\mathrm{rad}} > \frac{N(g)+N(v)}{D(g)+D(v)} \ge \frac{\delta_{n,\alpha}^{\mathrm{rad}} D(g) + \dfrac{(\xi_v+\gamma_{n,\alpha})^2}{\xi+h_{n,\alpha}} D(v)}{D(g)+D(v)} .$$

Noticing that $h_{n,\alpha}\delta_{n,\alpha}^{\mathrm{rad}} = \gamma_{n,\alpha}^2$, we infer that $\delta_{n,\alpha}^{\mathrm{rad}} - 2\gamma_{n,\alpha} > \xi_v \ge n-1$. Hence $\alpha < 2$ satisfies

$$3\alpha^2 - 2(n+4)\alpha - n^2 + 4n + 4 > 0 ,$$

that is,

$$\alpha < \frac{1}{3}\left(n+4-2\sqrt{n^2-n-1}\right) .$$

Conversely, if

$$\frac{1}{3}\left(n+4-2\sqrt{n^2-n-1}\right) \le \alpha < n,$$

then it necessarily holds that $\delta_{n,\alpha} = \delta_{n,\alpha}^{\mathrm{rad}}$. □

References

[1] Adimurthi, Grossi, M., Santra, S.: Optimal Hardy–Rellich inequalities, maximum principle and related eigenvalue problem. J. Funct. Anal. **240**, 36–83 (2006)

[2] Adimurthi, Santra, S.: Generalized Hardy–Rellich inequalities in critical dimensions and its applications. Commun. Contemp. Math. **11**, 367–394 (2009)

[3] Bennett, D.M.: An extension of Rellich's inequality. Proc. Amer. Math. Soc. **106**, 987–993 (1989)

[4] Berchio, E., Farina, A., Ferrero, A., Gazzola, F.: Existence and stability of entire solutions to a semilinear fourth order elliptic problem. J. Diff. Equations **252**, 2596–2616 (2012)

[5] Bhakta, M., Musina, R.: Entire solutions for a class of variational problems involving the biharmonic operator and Rellich potentials. Nonlinear Analysis T.M.A. **75**, 3836–3848 (2012)

[6] Caldiroli, P., Musina, R.: Rellich inequalities with weights. Calc. Var. **45**, 147–164 (2012)

[7] Caldiroli, P., Musina, R.: On Caffarelli–Kohn–Nirenberg type inequalities for the weighted biharmonic operator in cones. Milan J. Math. **79**, 657–687 (2011)

[8] Evans, W.D., Lewis, R.T.: On the Rellich inequality with magnetic potentials. Math. Z. **251**, 267–284 (2005)

[9] Gazzola, F., Grunau, H.-C., Mitidieri, E.: Hardy inequalities with optimal constants and remainder terms. T. Am. Math. Soc. **356**, 2149–2168 (2003)

[10] Gazzola, F., Grunau, H.-C., Sweers, G.: Optimal Sobolev and Hardy–Rellich constants under Navier boundary conditions. Ann. Mat. Pura Appl. (4) **189**, 475–486 (2010)

[11] Ghoussoub, N., Moradifam, A.: Bessel pairs and optimal Hardy and Hardy–Rellich inequalities. Math. Ann. **349**, 1–57 (2011)

[12] Mitidieri, E.: A simple approach to Hardy's inequalities. Math. Notes **67**, 479–486 (2000) (translation from Mat. Zametki **67**, 563–572 (2000))

[13] Moradifam, A.: Optimal weighted Hardy–Rellich inequalities on $H^2 \cap H^1_0$. J. London. Math. Soc. **85**, 22–40 (2011)

[14] Musina, R.: Weighted Sobolev spaces of radially symmetric functions. Ann. Mat. Pura Appl. (to appear) DOI: 10.1007/s10231-013-0348-4

[15] Rellich, F.: Halbbeschränkte Differentialoperatoren höherer Ordnung. In: Gerretsen, J.C.H., de Groot, J. (eds.) Proceedings of the International Congress of Mathematicians (1954), vol. III, pp. 243–250. Noordhoff, Groningen (1956)

[16] Rellich, F.: Perturbation Theory of Eigenvalue Problems. Gordon and Breach, New York (1969)

[17] Tertikas, A., Zographopoulos, N.B.: Best constants in the Hardy–Rellich inequalities and related improvements. Adv. Math. **209**, 407–459 (2007)

Paolo Caldiroli
Dipartimento di Matematica
Università di Torino
via Carlo Alberto, 10
I-10123 Torino, Italy
e-mail: `paolo.caldiroli@unito.it`

Roberta Musina
Dipartimento di Matematica ed Informatica
Università di Udine
via delle Scienze, 206
I-33100 Udine, Italy
e-mail: `roberta.musina@uniud.it`

Concentration Analysis and Applications to PDE
Trends in Mathematics, 29–47

Blow-up Solutions for Linear Perturbations of the Yamabe Equation

Pierpaolo Esposito, Angela Pistoia and Jérôme Vétois

Abstract. For a smooth, compact Riemannian manifold (M,g) of dimension $N \geq 3$, we are interested in the critical equation

$$\Delta_g u + \left(\frac{N-2}{4(N-1)} S_g + \varepsilon h\right) u = u^{\frac{N+2}{N-2}} \quad \text{in } M\,, \quad u > 0 \quad \text{in } M\,,$$

where Δ_g is the Laplace–Beltrami operator, S_g is the scalar curvature of (M,g), $h \in C^{0,\alpha}(M)$, and ε is a small parameter.

Mathematics Subject Classification (2010). Primary 35J61; Secondary 53C21, 35B33.

Keywords. Yamabe equation, conformal geometry, linear perturbations, blow-up.

1. Introduction

Letting (M,g) be a smooth, compact Riemannian N-manifold, $N \geq 3$, we consider the solutions $u \in C^{2,\alpha}$ of the problem

$$\Delta_g u + \kappa u = c u^p, \quad u > 0 \quad \text{in } M\,, \tag{1.1}$$

where $\Delta_g := -\operatorname{div}_g \nabla$ is the Laplace–Beltrami operator, $\kappa \in C^{0,\alpha}(M)$, $\alpha \in (0,1)$, $c \in \mathbb{R}$, and $p > 1$.

When $\kappa = \alpha_N S_g$ and $p = 2^* - 1$, where $\alpha_N := \frac{N-2}{4(N-1)}$, S_g is the scalar curvature of (M,g) and $2^* := \frac{2N}{N-2}$ is the critical Sobolev exponent, equation (1.1) reads as

$$\Delta_g u + \frac{N-2}{4(N-1)} S_g u = c u^{\frac{N+2}{N-2}}, \quad u > 0 \quad \text{in } M\,, \tag{1.2}$$

The first author was partially supported by the Prin project "Critical Point Theory and Perturbative Methods for Nonlinear Differential Equations" and the Firb-Ideas project "Analysis and Beyond". The third author was partially supported by the ANR grant ANR-08-BLAN-0335-01.

and is referred to in the literature as the Yamabe problem. The constant c can be restricted to the values $-1/1$ or 0 depending on whether the *Yamabe invariant* of (M, g), namely

$$\mu_g(M) = \inf_{\widetilde{g} \in [g]} \left(\mathrm{Vol}_{\widetilde{g}}(M)^{\frac{2-N}{N}} \int_M S_{\widetilde{g}} dv_{\widetilde{g}} \right)$$

has negative/positive sign or vanishes, respectively, where

$$[g] = \{\phi g : \phi \in C^\infty(M), \phi > 0\}$$

is the conformal class of g and $\mathrm{Vol}_{\widetilde{g}}(M)$ is the volume of the manifold $(M, \widetilde{g})$. If u is a solution of (1.2), then the metric $\widetilde{g} = u^{4/(N-2)} g$ has constant scalar curvature and belongs to $[g]$.

The Yamabe problem, raised by H. Yamabe [42] in '60, was firstly solved by Trudinger [41] when $\mu_g(M) \leq 0$. In this case, the solution is unique (up to a normalization when $\mu_g(M) = 0$). In general, a solution of (1.2) can be found by a direct constrained minimization method. As shown by Aubin [1], the inequality

$$\mu_g(M) < \mu_{g_0}(\mathbb{S}^N), \tag{1.3}$$

where $(\mathbb{S}^N, g_0)$ is the round sphere, is the key ingredient to show compactness of minimizing sequences, a non-trivial fact in view of the non-compactness of the Sobolev embedding $H_1^2(M) \hookrightarrow L^{2^*}(M)$. $(\mathbb{S}^N, g_0)$ has already constant scalar curvature. For manifolds (M, g) which are not conformally equivalent to $(\mathbb{S}^N, g_0)$ ($(M, g) \neq (\mathbb{S}^N, g_0)$ for short) with $\mu_g(M) > 0$, the Yamabe equation (1.2) has been solved via (1.3) by:

- Aubin [1] in the non-locally conformally flat case with $N \geq 6$, by exploiting the non-vanishing of the Weyl curvature tensor Weyl_g of (M, g) in the construction of local test functions;
- Schoen [37] when either $N = 3, 4, 5$ or $(M, g) \neq (\mathbb{S}^N, g_0)$ is locally conformally flat, by exploiting the Positive Mass Theorem by Schoen–Yau [39, 40] in the construction of global test functions

(see also Lee–Parker [22] for a unified approach).

From now on, we restrict our attention to the case where (M, g) has *positive Yamabe invariant* $\mu_g(M) > 0$. When $(M, g) \neq (\mathbb{S}^N, g_0)$, Schoen [38] addressed the question of the compactness of Yamabe metrics, and he proved the compactness to be true in the locally conformally flat case [38]. Recently, compactness of Yamabe metrics has been proved to be true for a general manifold $(M, g) \neq (\mathbb{S}^N, g_0)$ of dimension $N \leq 24$ by Khuri–Marques–Schoen [21]. Unexpectedly, compactness of Yamabe metrics has revealed to be false in general in dimensions $N \geq 25$ by Brendle [5] and Brendle–Marques [6]. Previous contributions where the compactness of Yamabe metrics is proved in lower dimensions are by Li–Zhu [27] ($N = 3$), Druet [10] ($N \leq 5$), Marques [28] ($N \leq 7$), and Li–Zhang [24, 25, 26] ($N \leq 11$). In all these results, it is shown that sequences of solutions $(u_k)_{k \in \mathbb{N}}$ of (1.1) with $\kappa \equiv \alpha_N S_g$, $c = 1$, and exponents $(p_k)_{k \in \mathbb{N}}$ in $[1 + \varepsilon_0, 2^* - 1]$, $\varepsilon_0 > 0$ fixed, are pre-compact in $C^{2,\alpha}(M)$, $\alpha \in (0, 1)$.

When $\kappa \not\equiv \alpha_N S_g$, the situation is different. When $\kappa < \alpha_N S_g$, Druet [9, 10] (see also Druet–Hebey [13] and Druet–Hebey–Vétois [16]) proved that compactness does hold for equation (1.1) with $c = 1$ and exponents p in the range $[1 + \varepsilon_0, 2^* - 1]$, for all dimensions $N \geq 3$ (in case $N = 3$, it is possible to write a more refined condition on the mass, see Li–Zhu [27]). As shown in Micheletti–Pistoia–Vétois [29] and Pistoia–Vétois [32], in dimensions $N \geq 4$, such a compactness result does not hold when $\kappa(\xi_0) > \alpha_n S_g(\xi_0)$ at some point $\xi_0 \in M$ with a nondegeneracy assumption at ξ_0, and, see [29], compactness does not hold either in the supercritical range $p > 2^* - 1$ when $\kappa(\xi_0) < \alpha_N S_g(\xi_0)$ at some point $\xi_0 \in M$. We also refer to Robert–Vétois [36, Theorem 2.3] where a special noncompactness result is obtained in dimension $N = 6$ for potentials $\kappa > \alpha_N S_g$ (see also Druet [9] and Druet–Hebey [11, 12] in case of $(M, g) = (\mathbb{S}^N, g_0)$ with $N = 6$). In the locally conformally flat case with $N \geq 4$, Hebey–Vaugon [19] proved that there always exists $\widetilde{g} \in [g]$ such that the equation $\Delta_{\widetilde{g}} u + \alpha_N \max_M(S_{\widetilde{g}})u = u^{2^*-1}$ in M is not compact. In case $(M, g) = (\mathbb{S}^N, g_0)$ with $N \geq 5$ and when $(\kappa - \alpha_N S_g)$ is a positive constant, Chen–Wei–Yan [8] proved that equation (1.1) with $c = 1$ and $p = 2^* - 1$ is not compact (see also the constructions by Hebey–Wei [20] in case $N = 3$).

When the potential κ varies, for manifolds $(M, g) \neq (\mathbb{S}^N, g_0)$ with $\mu_g(M) > 0$, Druet [10] (see also Druet–Hebey [14]) proved that sequences of solutions $(u_k)_{k\in\mathbb{N}}$ of (1.1) with $c = 1$, exponents $(p_k)_{k\in\mathbb{N}}$ in $[1 + \varepsilon_0, 2^* - 1]$, and potentials $(\kappa_k)_{k\in\mathbb{N}}$, are pre-compact in $C^{2,\alpha}(M)$, $\alpha \in (0, 1)$, when $n = 3, 4, 5$ provided that $\kappa_k \leq \alpha_n S_g$. The same result is strongly expected to be true in the locally conformally flat case and generally for $N \leq 24$.

The aim of the paper is to investigate the effect of positive perturbations of the geometric potential by exhibiting the failure of compactness properties for the equation

$$\Delta_g u + (\alpha_N S_g + \varepsilon h)u = u^{2^*-1}, \quad u > 0 \quad \text{in } M\,, \tag{1.4}$$

where $h \in C^{0,\alpha}(M)$, $\alpha \in (0, 1)$, with $\max_M h > 0$ and $\varepsilon > 0$ is a small parameter.

A family $(u_\varepsilon)_\varepsilon$ of solutions to equation (1.4) is said to *blow up* at some point $\xi_0 \in M$ if there holds $\sup_U u_\varepsilon \to +\infty$ as $\varepsilon \to 0$, for all neighborhoods U of ξ_0 in M. Letting

$$E(\xi) := \frac{h(\xi)}{\left|\mathrm{Weyl}_g(\xi)\right|_g}\,,$$

our main result is:

Theorem 1.1. *Let $(M, g) \neq (\mathbb{S}^N, g_0)$ be a smooth, compact, non-locally conformally flat Riemannian manifold with $N \geq 6$ and $\mu_g(M) > 0$. Let $h \in C^{0,\alpha}(M)$, $\alpha \in (0, 1)$, so that $\max_M h > 0$ and $\inf\{|\mathrm{Weyl}_g(x)|_g : h(x) > 0\} > 0$. Then for $\varepsilon > 0$ small, equation (1.4) has a solution u_ε such that the family $(u_\varepsilon)_\varepsilon$ blows up, up to a sub-sequence, as $\varepsilon \to 0$ at some point ξ_0 so that $E(\xi_0) = \max_M E$.*

Introducing the "reduced energy" $\widetilde{E} : (0, \infty) \times M \to \mathbb{R}$ defined as

$$\widetilde{E}(d, \xi) = c_2 d^2 h(\xi) - c_3 d^4 \left|\mathrm{Weyl}_g(\xi)\right|_g^2$$

with $c_2, c_3 > 0$, Theorem 1.1 is an easy consequence of the following more general result:

Theorem 1.2. *Let $(M, g) \neq (\mathbb{S}^N, g_0)$ be a smooth, compact, non-locally conformally flat Riemannian manifold with $N \geq 6$ and $\mu_g(M) > 0$, and $h \in C^{0,\alpha}(M)$, $\alpha \in (0, 1)$. Assume that there exists a C^0-stable critical set $\mathcal{D} \subset (0, \infty) \times M$ of $\widetilde{E}$. Then for $\varepsilon > 0$ small, equation (1.4) has a solution u_ε such that the family $(u_\varepsilon)_\varepsilon$ blows up, up to a sub-sequence, at some $\xi_0 \in \pi(\mathcal{D})$, where $\pi : (0, \infty) \times M \to M$ is the projection operator onto the second component.*

According to Li [23], we say that a compact set $\mathcal{D} \subset (0, \infty) \times M$ of critical points of $\widetilde{E}$ is a *C^0-stable critical set* of $\widetilde{E}$ if for any compact neighborhood U of $\mathcal{D}$ in $(0, \infty) \times M$, there exists $\delta > 0$ such that, if $\mathcal{J} \in C^1(U)$ and $\|\mathcal{J} - \widetilde{E}\|_{C^0(U)} \leq \delta$, then $\mathcal{J}$ has at least one critical point in U.

Given $\xi \in M$ so that $h(\xi) > 0$, define $d(\xi)$ as

$$d(\xi) = \left(\frac{c_2 h(\xi)}{2c_3 \left|\mathrm{Weyl}_g(\xi)\right|_g^2}\right)^{1/2}$$

with the convention that $d(\xi) = +\infty$ if $\mathrm{Weyl}_g(\xi) = 0$. Given $\xi \in M$ with $h(\xi) > 0$, the function $\widetilde{E}$ is increasing for $d \in (0, d(\xi))$ and, if $d(\xi) < +\infty$, achieves its global maximum in d at $d(\xi)$. Since

$$\widetilde{E}(d(\xi), \xi) = \frac{c_2^2 h^2(\xi)}{4c_3 \left|\mathrm{Weyl}_g(\xi)\right|_g^2} = \frac{c_2^2}{4c_3} E(\xi)^2,$$

in order to derive Theorem 1.1, the set $\mathcal{D}$ in Theorem 1.2 is constructed as

$$\mathcal{D} = \{(d(\xi), \xi) : \xi \in M \text{ s.t. } E(\xi) = \max_M E\},$$

which is clearly a *C^0-stable critical set* of $\widetilde{E}$. Since $d(\xi)$ is a maximum point of $\widetilde{E}$ in d, neither minimum points of E, nor saddle points of E can provide any *C^0-stable critical set* of $\widetilde{E}$.

Let us finally compare problem (1.4) with its Euclidean counter-part on a smooth bounded domain $\Omega \subset \mathbb{R}^N$, $N \geq 4$, with homogeneous Dirichlet boundary condition:

$$\Delta_{\mathrm{Eucl}} u + \lambda u = u^{2^*-1} \text{ in } \Omega, \quad u > 0 \text{ in } \Omega, \quad u = \text{ on } \partial\Omega. \tag{1.5}$$

For $\lambda \geq 0$, a direct minimization method (for the corresponding Rayleigh quotient) never gives rise to any solution of (1.5), and no solutions exist at all if Ω is star-shaped as shown by Pohožaev [33]. Moreover, following the arguments developed by Ben Ayed–El Mehdi–Grossi–Rey [3], problem (1.5) has never any solution with

a single blow-up point as $\lambda \to 0^+$. The effect of the geometry, which is crucial to provide a solution for the Yamabe problem (corresponding to $\lambda = 0$ in (1.5)) by minimization, is also relevant to producing solutions of (1.4) (corresponding to $\lambda \to 0^+$ in (1.5)) with a single blow-up point as stated in Theorem 1.1.
When $\lambda < 0$, solutions of (1.5) can be found by direct minimization as shown by Brezis–Nirenberg [7], and exhibit a single blow-up point as $\lambda \to 0^-$ as shown by Han [18], in contrast with the compactness property proved by Druet [9, 10]. Solutions of (1.5) with a single blow-up point, see Rey [35, 34], and with multiple blow-up points, see Bahri–Li–Rey [2] and Musso–Pistoia [30], as $\lambda \to 0^-$ have been constructed in a very general way.

We attack the existence issue of blowing-up solutions by a perturbative method, referred to in the literature as the non-linear Lyapunov–Schmidt reduction. Such a method is well known and the main point is to produce a suitable ansatz for the solutions. In the non-locally conformally flat case with $N \geq 6$ the basic ansatz is like in Aubin [1], but, see Section 2, needs to be slightly corrected via linearization so to account for the local geometry. A similar idea has been used for the prescribed Q-curvature problem by Pistoia–Vaira [31], the fourth-order analogue of the Yamabe problem. An alternative and more geometrical approach can be devised based on the conformal covariance of $\Delta_g + \alpha_N S_g$. The main point is to allow the metric g to vary in the conformal class so to gain flatness at each point $\xi \in M$, and this approach allows us, see Esposito–Pistoia–Vétois [17], to cover in an unified way also the remaining cases $N = 4, 5$ or (M, g) locally conformally flat with $N \geq 6$ (the case $N = 3$ is always excluded by the compactness result of Li–Zhu [27]). The aim of this paper is at the same time to advertise the general result contained in [17], and to provide a simpler and more intuitive proof in a special case. Thanks to the solvability theory of the linearized operator, we are led to study critical points of a finite-dimensional functional $\mathcal{J}_\varepsilon$, and a key step is to obtain in Section 3 an asymptotic expansion of $\mathcal{J}_\varepsilon$ by identifying the "reduced energy" $\widetilde{E}$ as the main order term. In Section 4, we describe the main steps of the non-linear Lyapunov–Schmidt reduction, and we deduce our general result Theorem 1.2.

2. The correcting term towards an improved ansatz

Letting

$$U(r) = \left(\frac{\sqrt{N(N-2)}}{1+r^2} \right)^{\frac{N-2}{2}}, \tag{2.1}$$

we aim to solve

$$\Delta V + pU^{p-1}V = \frac{1}{3} \sum_{i,j=1}^{N} R_{ij}(\xi) \frac{y^i y^j}{|y|} \partial_r U + \alpha_N S_g(\xi) U\,, \tag{2.2}$$

where $p = \frac{N+2}{N-2}$ and R_{ij} are the components of the Ricci tensor Ric_g of (M,g) in geodesic coordinates. Here, $\Delta = \sum_{i=1}^{N} \frac{\partial^2}{\partial y_i^2}$ is the Euclidean Laplacian with the standard sign convention, and $U(|y|)$ is the unique positive radial solution of $-\Delta U = U^p$ with $U(0) = \max_{\mathbb{R}^N} U = [N(N-2)]^{\frac{N-2}{4}}$.

Since $S_g(\xi) = \sum_{i=1}^{N} R_{ii}(\xi)$, a straightforward computation shows that

$$V(y) = [N(N-2)]^{\frac{N-2}{4}} \left(\frac{|y|^2+3}{12(1+|y|^2)^{\frac{N}{2}}} \sum_{i,j=1}^{N} R_{ij}(\xi) y^i y^j - \frac{S_g(\xi)}{24(N-1)} \frac{|y|^4+3}{(1+|y|^2)^{\frac{N}{2}}} \right) \tag{2.3}$$

is a solution of (2.2) as we were searching for.

Let $0 < r_0 < i_g(M)$, where $i_g(M)$ is the injectivity radius of (M,g). Take χ a smooth cutoff function such that $0 \le \chi \le 1$ in $\mathbb{R}$, $\chi \equiv 1$ in $[-r_0/2, r_0/2]$, and $\chi \equiv 0$ out of $[-r_0, r_0]$. For any point ξ in M and for any positive real number μ, we define the functions $\mathcal{U}_{\mu,\xi}$ and $\mathcal{V}_{\mu,\xi}$ on M by

$$\mathcal{U}_{\mu,\xi}(z) = \chi\left(d_g(z,\xi)\right) U_\mu\left(d_g(z,\xi)\right), \quad \mathcal{V}_{\mu,\xi}(z) = \chi\left(d_g(z,\xi)\right) V_\mu\left(\exp_\xi^{-1}(z)\right),$$

where d_g is the geodesic distance in (M,g) and $\exp_\xi^{-1}$ is the geodesic coordinate system. Here, U_μ and V_μ are defined as

$$U_\mu(x) = \mu^{-\frac{N-2}{2}} U\left(\frac{x}{\mu}\right), \quad V_\mu(x) = \mu^{-\frac{N-2}{2}} V\left(\frac{x}{\mu}\right),$$

obtained by scaling U and V in (2.1) and (2.3), respectively. Since $\mu_g(M) > 0$ implies the coercivity of the conformal Laplacian $\Delta_g + \alpha_N S_g$, let $i^* : L^{\frac{2N}{N+2}}(M) \to H_g^1(M)$ be the bounded operator defined as follows: the function $u = i^*(w)$ is the unique solution in $H_g^1(M)$ of the equation $\Delta_g u + \alpha_N S_g u = w$ in M. Problem (1.4) re-writes as

$$u = i^*\left[u_+^p - \varepsilon h u\right], \tag{2.4}$$

and we look for solutions of (2.4) in the form

$$u_\varepsilon(z) = \mathcal{W}_{\mu,\xi}(z) + \phi_\varepsilon(z), \quad \mathcal{W}_{\mu,\xi} = \mathcal{U}_{\mu,\xi} + \mu^2 \mathcal{V}_{\mu,\xi}, \tag{2.5}$$

where $\xi \in M$, $\mu > 0$ is small and ϕ_ε is a small remainder term.

First of all, we introduce the error term

$$\mathcal{R}_{\mu,\xi} = \mathcal{W}_{\mu,\xi} - i^*\left[(\mathcal{W}_{\mu,\xi})_+^p - \varepsilon h \mathcal{W}_{\mu,\xi}\right]. \tag{2.6}$$

We want to point out that the choice of the ansatz in (2.5) with the extra term $\mathcal{V}_{\mu,\xi}$ is motivated by the need that the error term has to be small enough. Indeed, the error term is estimated as follows.

Lemma 2.1. *Let $N \geq 6$. There exists a positive constant $C_0 > 0$ such that for any μ small and ξ in M there holds*

$$\|\mathcal{R}_{\mu,\xi}\| \leq C_0 \begin{cases} \mu^{\frac{N-2}{2}} + \varepsilon\mu^2 |\ln \mu|^{\frac{2}{3}} & \text{if } N = 6 \\ \mu^{\frac{N-2}{2}} + \varepsilon\mu^2 & \text{if } N = 7 \\ \mu^3 |\ln \mu|^{\frac{5}{8}} + \varepsilon\mu^2 & \text{if } N = 8 \\ \mu^3 + \varepsilon\mu^2 & \text{if } N = 9 \\ \mu^{2\frac{N+2}{N-2}} + \varepsilon\mu^2 & \text{if } N \geq 10. \end{cases} \tag{2.7}$$

Proof. It is enough to estimate the $L^{\frac{2N}{N+2}}$-norm of

$$\Delta_g \mathcal{W}_{\mu,\xi} + (\alpha_N S_g + \varepsilon h)\mathcal{W}_{\mu,\xi} - (\mathcal{W}_{\mu,\xi})_+^p .$$

Since $\mathcal{U}_{\mu,\xi} \circ \exp_\xi$ is radially symmetric in $B_0(r_0)$, we have that

$$\Delta_g \mathcal{U}_{\mu,\xi} (\exp_\xi x) = -\Delta \left(\mathcal{U}_{\mu,\xi} \circ \exp_\xi\right)(x) - \frac{1}{2}\partial_r (\ln |g|) \partial_r \left(\mathcal{U}_{\mu,\xi} \circ \exp_\xi\right)(x),$$

where $|g| := \det g$. In geodesic coordinates, we have the Taylor expansion

$$|g| = 1 - \frac{1}{3}\sum_{i,j=1}^{N} R_{ij}(\xi) x^i x^j + \mathrm{O}(|x|^3) \tag{2.8}$$

(see for example Lee–Parker [22]), yielding to

$$\begin{aligned} \Delta_g \mathcal{U}_{\mu,\xi} (\exp_\xi x) &= -\chi(|x|)\Delta U_\mu(x) + \frac{\chi(|x|)}{3}\sum_{i,j=1}^{N} \frac{R_{ij}(\xi) x^i x^j}{|x|}\partial_r U_\mu(x) \\ &\quad + \mathrm{O}\left(\mu^{\frac{N-2}{2}} + |x|^2 |\nabla U_\mu|\right) \\ &= \mathcal{U}^p_{\mu,\xi}(\exp_\xi x) + \frac{\chi(|x|)}{3}\sum_{i,j=1}^{N} \frac{R_{ij}(\xi) x^i x^j}{|x|}\partial_r U_\mu(x) \\ &\quad + \mathrm{O}\left(\mu^{\frac{N-2}{2}} + |x|^2 |\nabla U_\mu|\right) \end{aligned} \tag{2.9}$$

in view of $-\Delta U_\mu = U_\mu^p$. Similarly, we have that

$$\Delta_g \mathcal{V}_{\mu,\xi} (\exp_\xi x) = -\chi(|x|)\Delta V_\mu(x) + \mathrm{O}\left(\mu^{\frac{N-6}{2}} + |x||\nabla V_\mu|\right).$$

Since by (2.2) we have that

$$\Delta(\mu^2 V_\mu) + pU_\mu^{p-1}(\mu^2 V_\mu) = \frac{1}{3}\sum_{i,j=1}^{N} R_{ij}(\xi)\frac{x^i x^j}{|x|}\partial_r U_\mu + \alpha_N S_g(\xi) U_\mu, \tag{2.10}$$

by (2.9)–(2.10) we get that

$$\|\Delta_g \mathcal{W}_{\mu,\xi} + \alpha_N S_g \mathcal{W}_{\mu,\xi} - \mathcal{U}^p_{\mu,\xi} - p\mu^2 \mathcal{U}^{p-1}_{\mu,\xi} \mathcal{V}_{\mu,\xi}\|_{L^{\frac{2N}{N+2}}(M)}$$
$$= \begin{cases} \mathrm{O}\left(\mu^{\frac{N-2}{2}}\right) & \text{if } N = 6,7 \\ \mathrm{O}\left(\mu^3 |\ln \mu|^{\frac{5}{8}}\right) & \text{if } N = 8 \\ \mathrm{O}\left(\mu^3\right) & \text{if } N \geq 9. \end{cases} \tag{2.11}$$

Since

$$\|h\mathcal{W}_{\mu,\xi}\|_{L^{\frac{2N}{N+2}}(M)} = \begin{cases} \mathrm{O}\left(\mu^2 |\ln \mu|^{\frac{2}{3}}\right) & \text{if } N = 6 \\ \mathrm{O}\left(\mu^2\right) & \text{if } N \geq 7 \end{cases}$$

and

$$\| \left(\mathcal{W}_{\mu,\xi}\right)^p_+ - \mathcal{U}^p_{\mu,\xi} - p\, \mathcal{U}^{p-1}_{\mu,\xi} \left(\mu^2 \mathcal{V}_{\mu,\xi}\right) \|_{L^{\frac{2N}{N+2}}(M)} = \begin{cases} \mathrm{O}\left(\mu^4 |\ln \mu|^{\frac{2}{3}}\right) & \text{if } N = 6 \\ \mathrm{O}\left(\mu^{2\frac{N+2}{N-2}}\right) & \text{if } N \geq 7 \end{cases}$$

in view of $|(a+b)^p_+ - a^p - pa^{p-1}b| = \mathrm{O}(|b|^p)$ for all $a > 0$ and $b \in \mathbb{R}$, by (2.11) we deduce the validity of (2.7). □

3. The reduced energy

Introduce the Euler–Lagrange functional $J_\varepsilon : \mathrm{H}^1_g(M) \to \mathbb{R}$ corresponding to equation (1.4):

$$J_\varepsilon(u) := \frac{1}{2}\int_M |\nabla u|^2_g dv_g + \frac{1}{2}\int_M \left(\alpha_N S_g + \varepsilon h\right) u^2 dv_g - \frac{1}{p+1}\int_M u_+^{p+1} dv_g\,.$$

The aim is to find an asymptotic expansion of $J_\varepsilon\left(\mathcal{W}_{\mu,\xi}\right)$. We have that:

Proposition 3.1. *The following expansions do hold as* ϵ, $\mu \to 0$:

$$J_\varepsilon\left(\mathcal{W}_{\mu,\xi}\right) = \frac{K_6^{-6}}{6} + \frac{4}{5}\omega_5 \left|\mathrm{Weyl}_g(\xi)\right|^2_g \mu^4 \ln \mu + \frac{5}{24} K_6^{-6} h(\xi)\varepsilon\mu^2 + \mathrm{o}\left(\mu^4 \ln \mu + \varepsilon\mu^2\right) \tag{3.1}$$

when $N = 6$, *and*

$$J_\varepsilon\left(\mathcal{W}_{\mu,\xi}\right) = \frac{K_N^{-N}}{N} - \frac{K_N^{-N}}{24N(N-4)(N-6)} \left|\mathrm{Weyl}_g(\xi)\right|^2_g \mu^4 + \frac{2(N-1)K_N^{-N} h(\xi)}{N(N-2)(N-4)} \varepsilon\mu^2 + \mathrm{o}\left(\mu^4 + \varepsilon\mu^2\right) \tag{3.2}$$

when $N \geq 7$, *uniformly with respect to* $\xi \in M$, *where* K_N *is the best constant for the embedding of* $D^{1,2}\left(\mathbb{R}^N\right)$ *into* $L^{2^*}\left(\mathbb{R}^N\right)$.

Proof. Since we can write

$$\begin{aligned} &J_\varepsilon\left(\mathcal{U}_{\mu,\xi}+\mu^2\mathcal{V}_{\mu,\xi}\right)-J_\varepsilon\left(\mathcal{U}_{\mu,\xi}\right)\\ &=\mu^2\int_M\left[\langle\nabla\mathcal{U}_{\mu,\xi},\nabla\mathcal{V}_{\mu,\xi}\rangle_g+(\alpha_N S_g+\varepsilon h)\mathcal{U}_{\mu,\xi}\mathcal{V}_{\mu,\xi}-\mathcal{U}^p_{\mu,\xi}\mathcal{V}_{\mu,\xi}\right]\\ &\quad+\frac12\mu^4\int_M\left[|\nabla\mathcal{V}_{\mu,\xi}|^2_g-p\mathcal{U}^{p-1}_{\mu,\xi}\mathcal{V}^2_{\mu,\xi}\right]dv_g+\frac12\mu^4\int_M(\alpha_N S_g+\varepsilon h)\ \mathcal{V}^2_{\mu,\xi}dv_g\\ &\quad-\frac{1}{p+1}\int_M\left[\left(\mathcal{U}_{\mu,\xi}+\mu^2\mathcal{V}_{\mu,\xi}\right)^{p+1}_+-\mathcal{U}^{p+1}_{\mu,\xi}-(p+1)\mathcal{U}^p_{\mu,\xi}\mu^2\mathcal{V}_{\mu,\xi}\right.\\ &\quad\left.-\frac12 p(p+1)\mathcal{U}^{p-1}_{\mu,\xi}\mu^4\mathcal{V}^2_{\mu,\xi}\right]dv_g, \end{aligned}$$

first we have that

$$\begin{aligned} &J_\varepsilon\left(\mathcal{U}_{\mu,\xi}+\mu^2\mathcal{V}_{\mu,\xi}\right)-J_\varepsilon\left(\mathcal{U}_{\mu,\xi}\right)\\ &=\mu^2\int_M\left[\Delta_g\mathcal{U}_{\mu,\xi}+\alpha_N S_g\mathcal{U}_{\mu,\xi}-\mathcal{U}^p_{\mu,\xi}\right]\mathcal{V}_{\mu,\xi}dv_g\\ &\quad+\frac12\mu^4\int_M\left[\Delta_g\mathcal{V}_{\mu,\xi}-p\mathcal{U}^{p-1}_{\mu,\xi}\mathcal{V}_{\mu,\xi}\right]\mathcal{V}_{\mu,\xi}dv_g\\ &\quad+\begin{cases}\mathrm{o}\left(\mu^4|\ln\mu|+\varepsilon\mu^2\right) & \text{if } N=6\\ \mathrm{o}\left(\mu^4+\varepsilon\mu^2\right) & \text{if } N\ge 7\end{cases} \end{aligned}\tag{3.3}$$

as $\mu\to0$, in view of

$$\begin{aligned} &\int_M\left|\left(\mathcal{U}_{\mu,\xi}+\mu^2\mathcal{V}_{\mu,\xi}\right)^{p+1}_+-\mathcal{U}^{p+1}_{\mu,\xi}-(p+1)\mathcal{U}^p_{\mu,\xi}\mu^2\mathcal{V}_{\mu,\xi}-\frac12p(p+1)\mathcal{U}^{p-1}_{\mu,\xi}\mu^4\mathcal{V}^2_{\mu,\xi}\right|dv_g\\ &=\mathrm{O}\left(\mu^{\frac{4N}{N-2}}\int_M|\mathcal{V}_{\mu,\xi}|^{\frac{2N}{N-2}}dv_g\right)=\mathrm{o}\left(\mu^4\right) \end{aligned}$$

and

$$\int_M\mathcal{V}^2_{\mu,\xi}dv_g=\begin{cases}\mathrm{O}(1) & \text{if } N=6\\ \mathrm{o}(1) & \text{if } N\ge7\end{cases}$$

as $\mu\to0$. Now, observe that there holds

$$\begin{aligned} &\mu^2\int_M\left[\Delta_g\mathcal{U}_{\mu,\xi}+\alpha_N S_g\mathcal{U}_{\mu,\xi}-\mathcal{U}^p_{\mu,\xi}\right]\mathcal{V}_{\mu,\xi}dv_g+\mu^4\int_M\left[\Delta_g\mathcal{V}_{\mu,\xi}-p\mathcal{U}^{p-1}_{\mu,\xi}\mathcal{V}_{\mu,\xi}\right]\mathcal{V}_{\mu,\xi}dv_g\\ &=\mu^2\int_M\left[\Delta_g\mathcal{W}_{\mu,\xi}+\alpha_N S_g\mathcal{W}_{\mu,\xi}-\mathcal{U}^p_{\mu,\xi}-p\mu^2\mathcal{U}^{p-1}_{\mu,\xi}\mathcal{V}_{\mu,\xi}\right]\mathcal{V}_{\mu,\xi}dv_g\\ &\quad+\mathrm{O}\left(\mu^4\int_M\mathcal{V}^2_{\mu,\xi}dv_g\right)=\begin{cases}\mathrm{o}\left(\mu^4\ln\mu\right) & \text{if } N=6\\ \mathrm{o}\left(\mu^4\right) & \text{if } N\ge7\end{cases} \end{aligned}$$

as $\mu\to0$, in view of (2.11). By (2.8) and

$$\begin{aligned} \Delta_g\mathcal{V}_{\mu,\xi}(\exp_\xi x)=&-\Delta\left(\mathcal{V}_{\mu,\xi}\circ\exp_\xi\right)(x)\\ &+\mathrm{O}\left(|x|\left|\nabla\left(\mathcal{V}_{\mu,\xi}\circ\exp_\xi\right)(x)\right|+|x|^2\left|\nabla^2\left(\mathcal{V}_{\mu,\xi}\circ\exp_\xi\right)(x)\right|\right) \end{aligned}$$

we deduce that

$$\begin{aligned}&\int_M \left[\Delta_g \mathcal{V}_{\mu,\xi} - p\mathcal{U}_{\mu,\xi}^{p-1}\mathcal{V}_{\mu,\xi}\right]\mathcal{V}_{\mu,\xi}dv_g\\&\quad = -\int_{B_0\left(\frac{r_0}{2\mu}\right)}\left(\Delta V + pU^{p-1}V\right)V dy + \begin{cases}\mathrm{O}(1) & \text{if } N=6\\ \mathrm{o}(1) & \text{if } N\ge 7\end{cases}\end{aligned} \tag{3.4}$$

as $\mu\to 0$. By (3.3) and (3.4), we get that

$$\begin{aligned}J_\varepsilon\left(\mathcal{U}_{\mu,\xi}+\mu^2\mathcal{V}_{\mu,\xi}\right) &= J_\varepsilon\left(\mathcal{U}_{\mu,\xi}\right) + \frac{1}{2}\mu^4\int_{B_0\left(\frac{r_0}{2\mu}\right)}\left(\Delta V + pU^{p-1}V\right)V dy\\&\quad + \begin{cases}\mathrm{o}\left(\mu^4\ln\mu\right) & \text{if } N=6\\ \mathrm{o}\left(\mu^4\right) & \text{if } N\ge 7\end{cases}\end{aligned} \tag{3.5}$$

as $\mu\to 0$. By (2.2)–(2.3) and easy symmetry properties we deduce that

$$\begin{aligned}&\int_{B_0\left(\frac{r_0}{2\mu}\right)}\left(\Delta V + pU^{p-1}V\right)V dy\\&= -\frac{[N(N-2)]^{\frac{N-2}{2}}(N-2)}{36}\int_{B_0\left(\frac{r_0}{2\mu}\right)}\left(\sum_{i,j=1}^N R_{ij}(\xi)y^iy^j\right)^2\frac{|y|^2+3}{(1+|y|^2)^N}dy\\&\quad + \frac{[N(N-2)]^{\frac{N-2}{2}}\alpha_N}{72N(N-1)}S_g^2(\xi)\int_{B_0\left(\frac{r_0}{2\mu}\right)}(1+|y|^2)^{-N}\left[(7N-10)|y|^6\right.\\&\quad \left.+3(7N-8)|y|^4+3(7N-10)|y|^2-9N\right]dy\\&= -\frac{[N(N-2)]^{\frac{N-2}{2}}(N-2)}{36}\sum_{i,j,k,s=1}^N E_{ij}(\xi)E_{ks}(\xi)\int_{B_0\left(\frac{r_0}{2\mu}\right)}\frac{(|y|^2+3)y^iy^jy^ky^s}{(1+|y|^2)^N}dy\\&\quad -\omega_{N-1}\frac{[N(N-2)]^{\frac{N-2}{2}}(N-2)}{576N^2(N-1)^2}S_g^2(\xi)\left[(N-2)(N-4)I_N^{\frac{N+4}{2}}\right.\\&\quad \left.+3(N^2-8N+8)I_N^{\frac{N+2}{2}} - 3N(7N-10)I_N^{\frac{N}{2}} + 9N^2 I_N^{\frac{N-2}{2}}\right] + \mathrm{o}(1)\end{aligned} \tag{3.6}$$

as $\mu\to 0$, where the E_{ij}'s are the components of the traceless part $\mathrm{E}_g = \mathrm{Ric}_g - \frac{S_g}{N}g$ of the Ricci curvature Ric_g of (M,g) in geodesic coordinates and

$$I_p^q = \begin{cases}\displaystyle\int_0^{+\infty}\frac{r^q}{(1+r)^p}dr & \text{if } p-q>1\\ \displaystyle\int_0^{\frac{r_0^2}{4\mu^2}}\frac{r^q}{(1+r)^p}dr & \text{if } p-q\le 1.\end{cases}$$

Since integration by parts yields

$$I_{p+1}^q = \frac{p-q-1}{p}I_p^q \quad \text{and} \quad I_{p+1}^{q+1} = \frac{q+1}{p-q-1}I_{p+1}^q \tag{3.7}$$

as soon as $p - q > 1$, we have that

$$I_N^{\frac{N}{2}} = \frac{N}{N-2} I_N^{\frac{N-2}{2}} = \frac{N-4}{N+2} I_N^{\frac{N+2}{2}}$$
$$I_N^{\frac{N+4}{2}} = \begin{cases} -2\ln\mu + O(1) & \text{if } N = 6 \\ \frac{(N+2)(N+4)}{(N-4)(N-6)} I_N^{\frac{N}{2}} & \text{if } N \geq 7 \end{cases} \tag{3.8}$$

as $\mu \to 0$, and it can be easily checked that

$$I_N^{\frac{N}{2}} = \frac{N\omega_N}{2^{N-1}(N-2)\,\omega_{N-1}} = \frac{2K_N^{-N}}{[N(N-2)]^{\frac{N-2}{2}}(N-2)^2\omega_{N-1}} \tag{3.9}$$

(see Aubin [1]). Since for all $i \neq j$ there holds

$$\int_{S^{N-1}} (y^i)^4 dv_{g_0} = 3\int_{S^{N-1}} (y^i)^2 (y^j)^2 dv_{g_0} = \frac{3}{N(N+2)} \int_{S^{N-1}} |y|^4 dv_{g_0},$$

by (3.6) and (3.8)–(3.9) we deduce that

$$\int_{B_0(\frac{r_0}{2\mu})} \left(\Delta V + pU^{p-1}V\right) V dy = \frac{8}{3}\omega_5 |E_g(\xi)|_g^2 \ln\mu + \frac{16}{225}\omega_5 S_g^2(\xi)\ln\mu + O(1) \tag{3.10}$$

if $N = 6$, and

$$\begin{aligned} &\int_{B_0(\frac{r_0}{2\mu})} \left(\Delta V + pU^{p-1}V\right) V dy \\ &= -\frac{2N-7}{9N(N-2)(N-4)(N-6)} K_N^{-N} |E_g(\xi)|_g^2 \\ &\quad + \frac{(N-2)(N-7)}{36N^2(N-1)(N-4)(N-6)} K_N^{-N} S_g^2(\xi) + o(1) \end{aligned} \tag{3.11}$$

if $N \geq 7$. Inserting (3.10)–(3.11) into (3.5), by Lemma 3.2 below we deduce the validity of (3.1)–(3.2). □

We are left with proving the following:

Lemma 3.2. *The following expansions do hold as ϵ, $\mu \to 0$:*

$$\begin{aligned} J_\varepsilon(U_{\mu,\xi}) = \frac{K_6^{-6}}{6} &+ \left[\frac{4}{5}\left|\mathrm{Weyl}_g(\xi)\right|_g^2 - \frac{4}{3}|E_g(\xi)|_g^2 - \frac{8}{225} S_g^2(\xi)\right] \omega_5 \mu^4 \ln\mu \\ &+ \frac{5}{24} K_6^{-6} h(\xi)\varepsilon\mu^2 + o\left(\mu^4 \ln\mu + \varepsilon\mu^2\right) \end{aligned}$$

when $N = 6$, *and*

$$J_\varepsilon(U_{\mu,\xi}) = \frac{K_N^{-N}}{N} + \Bigg[-\frac{K_N^{-N}}{24N(N-4)(N-6)}\left|\mathrm{Weyl}_g(\xi)\right|_g^2 + \frac{(2N-7)K_N^{-N}}{18N(N-2)(N-4)(N-6)}\left|\mathrm{E}_g(\xi)\right|_g^2 - \frac{(N-2)(N-7)K_N^{-N}}{72N^2(N-1)(N-4)(N-6)}S_g(\xi)^2\Bigg]\mu^4 + \frac{2(N-1)K_N^{-N}}{N(N-2)(N-4)}h(\xi)\varepsilon\mu^2 + \mathrm{o}\left(\mu^4 + \varepsilon\mu^2\right)$$

when $N \geq 7$, *uniformly with respect to* $\xi \in M$.

Proof. There hold

$$\frac{1}{\omega_{N-1}r^{N-1}}\int_{\partial B_\xi(r)} h d\sigma_g = h(\xi) + \mathrm{O}(r), \tag{3.12}$$

$$\frac{1}{\omega_{N-1}r^{N-1}}\int_{\partial B_\xi(r)} S_g d\sigma_g = S_g(\xi) - \frac{1}{2N}\Lambda_g(\xi)r^2 + \mathrm{O}\left(r^4\right), \tag{3.13}$$

$$\frac{1}{\omega_{N-1}r^{N-1}}\int_{\partial B_\xi(r)} d\sigma_g = 1 - \frac{1}{6N}S_g(\xi)r^2 + A_g(\xi)r^4 + \mathrm{O}\left(r^5\right), \tag{3.14}$$

as $r \to 0$, uniformly with respect to ξ, where $d\sigma_g$ is the volume element of $\partial B_\xi(r)$, ω_{N-1} is the volume of the unit $(N-1)$–sphere, and where (see (3.17)–(3.18))

$$\Lambda_g(\xi) = \Delta_g S_g(\xi) + \frac{1}{3}S_g(\xi)^2 \tag{3.15}$$

and

$$A_g(\xi) = \frac{18\Delta_g S_g(\xi) + 8\left|\mathrm{Ric}_g(\xi)\right|_g^2 - 3\left|\mathrm{Rm}_g(\xi)\right|_g^2 + 5S_g(\xi)^2}{360N(N+2)}. \tag{3.16}$$

The orthogonal decomposition of Riemann curvature is given by

$$\left|\mathrm{Rm}_g(\xi)\right|_g^2 = \left|\mathrm{Weyl}_g(\xi)\right|_g^2 + \frac{4}{N-2}\left|\mathrm{E}_g(\xi)\right|_g^2 + \frac{2}{N(N-1)}S_g(\xi)^2, \tag{3.17}$$

where Weyl_g is the Weyl curvature of g and $\mathrm{E}_g = \mathrm{Ric}_g - \frac{S_g}{N}g$ is the traceless part of the Ricci curvature of g. Moreover, we get

$$\left|\mathrm{Ric}_g(\xi)\right|_g^2 = \left|\mathrm{E}_g(\xi)\right|_g^2 + \frac{1}{N}S_g(\xi)^2. \tag{3.18}$$

By (3.8) and (3.14), we compute

$$\int_M \left|\nabla U_{\mu,\xi}\right|_g^2 dv_g = [N(N-2)]^{\frac{N-2}{2}}(N-2)^2\int_0^{\frac{r_0}{2}}\frac{\mu^{N-2}r^2}{\left(\mu^2+r^2\right)^N}\int_{\partial B_\xi(r)} d\sigma_g dr + \mathrm{O}\left(\mu^{N-2}\right)$$

$$= [N(N-2)]^{\frac{N-2}{2}}(N-2)^2\omega_{N-1}$$
$$\times \int_0^{\frac{r_0}{2\mu}} \frac{r^{N+1}}{(1+r^2)^N}\left(1-\frac{1}{6N}S_g(\xi)\mu^2r^2 + A_g(\xi)\mu^4r^4 + \mathrm{O}\left(\mu^5r^5\right)\right)dr$$
$$+\mathrm{O}\left(\mu^{N-2}\right)$$
$$= \frac{[N(N-2)]^{\frac{N-2}{2}}(N-2)^2}{2}\omega_{N-1}$$
$$\times\left(I_N^{\frac{N}{2}} - \frac{1}{6N}I_N^{\frac{N+2}{2}}S_g(\xi)\mu^2 + I_N^{\frac{N+4}{2}}A_g(\xi)\mu^4 + \mathrm{O}\left(I_N^{\frac{N+5}{2}}\mu^5 + \mu^{N-2}\right)\right),$$

and then

$$\int_M |\nabla U_{\mu,\xi}|_g^2\,dv_g \tag{3.19}$$
$$= K_N^{-N}\left(1-\frac{N+2}{6N(N-4)}S_g(\xi)\mu^2\right) - 9216\,\omega_5 A_g(\xi)\mu^4\ln\mu + \mathrm{O}\left(\mu^4\right)$$

when $N=6$ and

$$\int_M |\nabla U_{\mu,\xi}|_g^2\,dv_g \tag{3.20}$$
$$= K_N^{-N}\left(1-\frac{N+2}{6N(N-4)}S_g(\xi)\mu^2 + \frac{(N+2)(N+4)}{(N-4)(N-6)}A_g(\xi)\mu^4\right) + \mathrm{O}\left(\mu^5\right)$$

when $N \geq 7$, in view of (3.9). Since by (3.7) there hold

$$I_{N-2}^{\frac{N-2}{2}} = \frac{4(N-1)(N-2)}{N(N-4)}I_N^{\frac{N}{2}} \quad \text{and} \quad I_{N-2}^{\frac{N}{2}} = \begin{cases} -2\ln\mu + \mathrm{O}(1) & \text{if } N=6 \\ \frac{4(N-1)(N-2)}{(N-4)(N-6)}I_N^{\frac{N}{2}} & \text{if } N\geq 7 \end{cases}$$

as $\mu \to 0$, by (3.13) we compute

$$\int_M S_g U_{\mu,\xi}^2\,dv_g \tag{3.21}$$
$$= [N(N-2)]^{\frac{N-2}{2}}\int_0^{\frac{r_0}{2}} \frac{\mu^{N-2}}{(\mu^2+r^2)^{N-2}}\int_{\partial B_\xi(r)} S_g\,d\sigma_g\,dr + \mathrm{O}\left(\mu^{N-2}\right)$$
$$= [N(N-2)]^{\frac{N-2}{2}}\omega_{N-1}\mu^2\int_0^{\frac{r_0}{2\mu}} \frac{r^{N-1}}{(1+r^2)^{N-2}}\left(S_g(\xi) - \frac{1}{2N}\Lambda_g(\xi)\mu^2r^2\right.$$
$$\left.+\mathrm{O}\left(\mu^4r^4\right)\right)dr + \mathrm{O}\left(\mu^{N-2}\right)$$
$$= \frac{[N(N-2)]^{\frac{N-2}{2}}}{2}\omega_{N-1}\mu^2\left(I_{N-2}^{\frac{N-2}{2}}S_g(\xi) - \frac{1}{2N}I_{N-2}^{\frac{N}{2}}\Lambda_g(\xi)\mu^2\right)$$
$$+\mathrm{O}\left(\mu^6 I_{N-2}^{\frac{N+2}{2}} + \mu^N\right)$$

$$= \begin{cases} \dfrac{5K_6^{-6}}{12}\mu^2 S_g(\xi) + 48\omega_5 A_g(\xi)\mu^4 \ln\mu + \mathrm{O}(\mu^4) & \text{if } N = 6 \\ \dfrac{4(N-1)K_N^{-N}}{N(N-2)(N-4)}\mu^2 \left(S_g(\xi) - \dfrac{A_g(\xi)\mu^2}{2(N-6)}\right) + \mathrm{O}(\mu^5) & \text{if } N \geq 7 \end{cases}$$

in view of (3.9). Similarly, by (3.12), we have that

$$\varepsilon \int_M h U_{\mu,\xi}^2 dv_g = \frac{4(N-1)K_N^{-N}}{N(N-2)(N-4)} h(\xi)\varepsilon\mu^2 + \mathrm{o}(\varepsilon\mu^2). \tag{3.22}$$

By (3.8) and (3.14), we compute

$$\int_M U_{\mu,\xi}^{2^*} dv_g = [N(N-2)]^{\frac{N}{2}} \int_0^{\frac{r_0}{2}} \frac{\mu^N \int_{\partial B_\xi(r)} d\sigma_g}{(\mu^2+r^2)^N} dr + \mathrm{O}(\mu^N) \tag{3.23}$$

$$= [N(N-2)]^{\frac{N}{2}} \omega_{N-1} \int_0^{\frac{r_0}{2\mu}} \frac{r^{N-1}}{(1+r^2)^N} \left(1 - \frac{S_g(\xi)}{6N}\mu^2 r^2 + A_g(\xi)\mu^4 r^4\right) dr + \mathrm{O}(\mu^5)$$

$$= \frac{[N(N-2)]^{\frac{N}{2}}}{2}\omega_{N-1}\left(I_N^{\frac{N-2}{2}} - \frac{I_N^{\frac{N}{2}}}{6N} S_g(\xi)\mu^2 + I_N^{\frac{N+2}{2}} A_g(\xi)\mu^4\right) + \mathrm{O}(\mu^5)$$

$$= K_N^{-N}\left(1 - \frac{S_g(\xi)\mu^2}{6(N-2)} + \frac{N(N+2)}{(N-2)(N-4)} A_g(\xi)\mu^4\right) + \mathrm{O}(\mu^5)$$

in view of (3.9). Finally, the claimed expansions follow by (3.19), (3.20), (3.21), (3.22) and (3.23) in view of (3.15)–(3.18). □

4. The Lyapunov–Schmidt reduction argument

Since equation (1.4) can be re-written as (2.4), the function $u = \mathcal{W}_{\mu,\xi} + \phi$ does solve (1.4) as soon as

$$\hat{L}_{\mu,\xi}(\phi) = -\mathcal{R}_{\mu,\xi} - N_{\mu,\xi}(\phi), \tag{4.1}$$

where $\mathcal{R}_{\mu,\xi}$ is given in (2.6),

$$N_{\mu,\xi}(\phi) = -i^* \left[(\mathcal{W}_{\mu,\xi}+\phi)_+^p - (\mathcal{W}_{\mu,\xi})_+^p - p(\mathcal{W}_{\mu,\xi})_+^{p-1}\phi\right]$$

is the nonlinear term (quadratic in ϕ) and

$$\begin{array}{rccl} \hat{L}_{\mu,\xi}: & H_g^1(M) & \to & H_g^1(M) \\ & \phi & \mapsto & \phi - i^*\left[p(\mathcal{W}_{\mu,\xi})_+^{p-1}\phi - \varepsilon h\phi\right] \end{array}$$

is the linearized operator of (2.4) at $\mathcal{W}_{\mu,\xi}$.

Since $\mathcal{W}_{\mu,\xi}$ is a small perturbation of $\mathcal{U}_{\mu,\xi}$, as $\varepsilon, \mu \to 0$ the operator $\hat{L}_{\mu,\xi}$ in balls with radii of order μ looks pretty much as a scaling of the limiting operator $L_\infty: \Phi \to \Phi + (\Delta)^{-1}\left[pU^{p-1}\Phi\right]$, where U is given in (2.1). It is well known (see Bianchi–Egnell [4]) that

$$\ker L_\infty = \text{Span}\left\{\Phi^0, \Phi^1, \ldots, \Phi^N\right\},$$

where

$$\Phi^0(y) = \frac{1-|y|^2}{(1+|y|^2)^{\frac{N}{2}}}, \qquad \Phi^i(y) = \frac{y^i}{(1+|y|^2)^{\frac{N}{2}}} \quad \forall\, i = 1,\dots,N. \tag{4.2}$$

Since there is no hope for the full invertibility of $\hat{L}_{\mu,\xi}$ in $H_g^1(M)$, let us introduce the "asymptotic kernel" $K_{\mu,\xi}$ and its "orthogonal space" $K_{\mu,\xi}^{\perp}$ as

$$K_{\mu,\xi} = \text{Span}\ \{Z_{\mu,\xi}^0,\dots,Z_{\mu,\xi}^N\}$$

and

$$K_{\mu,\xi}^{\perp} = \left\{\phi \in H_g^1(M) : \int_M \mathcal{U}_{\mu,\xi}^{p-1} Z_{\mu,\xi}^i \phi d\mu_g = 0 \quad \forall\, i = 0,\dots,N\right\},$$

where

$$Z_{\mu,\xi}^i(z) = \chi(d_g(z,\xi))\, \mu^{\frac{2-N}{2}} \Phi^i\left(\frac{\exp_\xi^{-1}(z)}{\mu}\right)$$

for $i = 0,\dots,N$, with Φ^i given by (4.2). Letting $\Pi_{\mu,\xi}$ and $\Pi_{\mu,\xi}^{\perp}$ be the projectors of $H_g^1(M)$ onto the respective subspaces, equation (4.1) is equivalent to solving

$$L_{\mu,\xi}(\phi) = -\Pi_{\mu\xi}^{\perp}\left(\mathcal{R}_{\mu,\xi} + N_{\mu,\xi}(\phi)\right), \tag{4.3}$$

$$\Pi_{\mu_\xi}\left(\hat{L}_{\mu,\xi}(\phi)\right) = -\Pi_{\mu_\xi}\left(\mathcal{R}_{\mu,\xi} + N_{\mu,\xi}(\phi)\right) \tag{4.4}$$

for some $\phi \in K_{\mu,\xi}^{\perp}$, where $L_{\mu,\xi} = \Pi_{\mu,\xi}^{\perp} \circ \hat{L}_{\mu,\xi} : K_{\mu,\xi}^{\perp} \to K_{\mu,\xi}^{\perp}$. First we can solve equation (4.3), a rather standard result in this context (see for example Musso–Pistoia [30]):

Lemma 4.1. *There exists a positive constant C_0 such that, for any ε,μ small and any $\xi \in M$, there holds*

$$\|L_{\mu,\xi}(\phi)\| \geq C_0 \|\phi\|$$

for all $\phi \in K_{\mu,\xi}^{\perp}$. As a consequence, (4.3) admits a unique solution $\phi_{\mu,\xi} \in K_{\mu,\xi}^{\perp}$, which is continuously differentiable in μ and ξ, so that

$$\|\phi_{\mu,\xi}\| = \begin{cases} \mathrm{o}\left(\mu^2\sqrt{|\ln\mu|} + \sqrt{\varepsilon}\mu\right) & \text{if } N = 6 \\ \mathrm{o}\left(\mu^2 + \sqrt{\varepsilon}\mu\right) & \text{if } N \geq 7. \end{cases} \tag{4.5}$$

Let us just stress out that the estimate (4.5) heavily depends on (2.7). The need of improving the ansatz in Section 2 comes out from getting the correct smallness rate of ϕ as expressed by (4.5). Finally, we have all the ingredients to prove our main result.

Proof of Theorem 1.2. A first well-known fact (see for example Musso–Pistoia [30]) is the equivalence between equation (4.4) and the search of critical points for

$$\widetilde{\mathcal{J}}_\varepsilon(\mu,\xi) = J_\varepsilon\left(\mathcal{W}_{\mu,\xi} + \phi_{\mu,\xi}\right),$$

where $\phi_{\mu,\xi}$ is given by Lemma 4.1. We just need to prove that

$$J_\varepsilon\left(\mathcal{W}_{\mu,\xi}+\phi_{\mu,\xi}\right)-J_\varepsilon\left(\mathcal{W}_{\mu,\xi}\right)=\begin{cases}\mathrm{o}\left(\mu^4|\ln\mu|+\varepsilon\mu^2\right) & \text{if } N=6\\ \mathrm{o}\left(\mu^4+\varepsilon\mu^2\right) & \text{if } N\geq 7\end{cases}\tag{4.6}$$

as $\varepsilon,\mu\to 0$. Indeed, we have that

$$\begin{aligned}
&J_\varepsilon\left(\mathcal{W}_{\mu,\xi}+\phi_{\mu,\xi}\right)-J_\varepsilon\left(\mathcal{W}_{\mu,\xi}\right)\\
&=\int_M\left(\left\langle\nabla\mathcal{R}_{\mu,\xi},\nabla\phi_{\mu,\xi}\right\rangle_g+\alpha_N S_g\mathcal{R}_{\mu,\xi}\phi_{\mu,\xi}-\left(W_{\mu,\xi}\right)_+^p\phi_{\mu,\xi}\right)dv_g\\
&\quad+\frac{1}{2}\int_M\left|\nabla\phi_{\mu,\xi}\right|_g^2dv_g+\frac{1}{2}\int_M\left(\alpha_N S_g+\varepsilon h\right)\phi_{\mu,\xi}^2dv_g\\
&\quad-\frac{1}{p+1}\int_M\left[\left(\mathcal{W}_{\mu,\xi}+\phi_{\mu,\xi}\right)_+^{p+1}-\left(\mathcal{W}_{\mu,\xi}\right)_+^{p+1}-(p+1)\left(\mathcal{W}_{\mu,\xi}\right)_+^p\phi_{\mu,\xi}\right]dv_g\\
&=\mathrm{O}\left(\left\|\mathcal{R}_{\mu,\xi}\right\|\left\|\phi_{\mu,\xi}\right\|+\left\|\phi_{\mu,\xi}\right\|^2+\int_M\left(\mathcal{W}_{\mu,\xi}\right)_+^{p-1}\phi_{\mu,\xi}^2dv_g+\int_M\phi_{\mu,\xi}^{p+1}dv_g\right)\\
&=\mathrm{O}\left(\left\|\mathcal{R}_{\mu,\xi}\right\|\left\|\phi_{\mu,\xi}\right\|+\left\|\phi_{\mu,\xi}\right\|^2\right)
\end{aligned}$$

by the Sobolev embedding $H_g^1(M)\hookrightarrow L^{p+1}(M)$ and Hölder's inequality. By (2.7) and (4.5) we then deduce the validity of (4.6). Setting

$$\mu(d)=d\begin{cases}l^{-1}(\varepsilon) & \text{if } N=6\\ \sqrt{\varepsilon} & \text{if } N\geq 7,\end{cases}$$

where $l:(0,e^{-\frac{1}{2}})\to(0,\frac{e^{-1}}{2})$ is defined as $l(\mu)=-\mu^2\ln\mu$, by Proposition 3.1 and (4.6) we deduce the following asymptotic estimates:

$$\mathcal{J}(d,\xi):=\frac{\widetilde{\mathcal{J}}_\varepsilon(\mu(d),\xi)-K_N^{-N}}{\varepsilon^2}\left(\ln l^{-1}(\varepsilon)\right)^\gamma=c_2d^2h(\xi)-c_3d^4\left|\mathrm{Weyl}_g(\xi)\right|_g^2+\mathrm{o}(1)$$

as $\varepsilon\to 0$, uniformly with respect to $\xi\in M$ and to d in compact subsets of $(0,\infty)$, where $c_2,c_3>0$ are suitable constants, $\gamma=1$ when $N=6$ and $\gamma=0$ when $N\geq 7$. Letting $\mathcal{D}\subset(0,\infty)\times M$ be a C^0-*stable critical set* of $\widetilde{E}$ and U be a compact neighborhood of $\mathcal{D}$ in $(0,\infty)\times M$, by the definition of stability it follows that $\mathcal{J}$ has a critical point $(d_\varepsilon,\xi_\varepsilon)\in U\subset(0,\infty)\times M$, for ε small. Up to a subsequence and taking U smaller and smaller, we can assume that $(d_\varepsilon,\xi_\varepsilon)\to(t_0,\xi_0)$ as $\varepsilon\to 0$ with $\xi_0\in\pi(\mathcal{D})$. By elliptic regularity theory $u_\varepsilon=\mathcal{W}_{\mu(d_\varepsilon),\xi_\varepsilon}+\phi_{\mu(d_\varepsilon),\xi_\varepsilon}$ is a solution of (1.4). Since $\xi_\varepsilon\to\xi_0$ and $\left\|\phi_{\mu(d_\varepsilon),\xi_\varepsilon}\right\|\to 0$ as $\varepsilon\to 0$, it is easily seen that $u_\varepsilon>0$ and $u_\varepsilon^{2^*}\rightharpoonup K_N^{-N}\delta_{\xi_0}$ in the measures sense as $\varepsilon\to 0$ (see for example Rey [35]), where δ_ξ denotes the Dirac mass measure at ξ. From very basic facts concerning the asymptotic analysis of solutions of Yamabe-type equations (see for example Druet–Hebey [12] and Druet–Hebey–Robert[15]), we get that the family $(u_\varepsilon)_\varepsilon$ blows up at the point ξ_0 as $\varepsilon\to 0$. □

References

[1] T. Aubin, *Equations différentielles non linéaires et problème de Yamabe concernant la courbure scalaire.* J. Math. Pures Appl. (9) **55** (1976), 269–296.

[2] A. Bahri, Y.-Y. Li, O. Rey, *On a variational problem with lack of compactness: the topological effect of the critical points at infinity.* Calc. Var. Partial Differential Equations **3** (1995), 67–93.

[3] M. Ben Ayed, K. El Mehdi, O. Rey, M. Grossi, *A nonexistence result of single peaked solutions to a supercritical nonlinear problem.* Commun. Contemp. Math. **5** (2003), 179–195.

[4] G. Bianchi, H. Egnell, *A note on the Sobolev inequality.* J. Funct. Anal. **100** (1991), 18–24.

[5] S. Brendle, *Blow-up phenomena for the Yamabe equation.* J. Amer. Math. Soc. **21**, 951–979.

[6] S. Brendle, F.C. Marques, *Blow-up phenomena for the Yamabe equation. II.* J. Differential Geom. **81** (2009), 225–250.

[7] H. Brézis, L. Nirenberg, *Positive solutions of nonlinear elliptic equations involving critical Sobolev exponents.* Comm. Pure Appl. Math. **36** (1983), 437–477.

[8] W.-Y. Chen, J. Wei, S. Yan, *Infinitely many solutions for the Schrödinger equations in $\mathbb{R}^n$ with critical growth.* J. Differential Equations **252** (2012), 2425–2447.

[9] O. Druet, *From one bubble to several bubbles: the low-dimensional case.* J. Differential Geom. **63** (2003), 399–473.

[10] O. Druet, *Compactness for Yamabe metrics in low dimensions.* Int. Math. Res. Not. **23** (2004), 1143–1191.

[11] O. Druet, E. Hebey, *Blow-up examples for second order elliptic PDEs of critical Sobolev growth.* Trans. Amer. Math. Soc. **357** (2005), 1915–1929.

[12] O. Druet, E. Hebey, *Elliptic equations of Yamabe type.* Int. Math. Res. Surv. **1** (2005), 1–113.

[13] O. Druet, E. Hebey, *Stability for strongly coupled critical elliptic systems in a fully inhomogeneous medium.* Anal. PDE **2** (2009), 305–359.

[14] O. Druet, E. Hebey, *Stability and instability for Einstein-scalar field Lichnerowicz equations on compact Riemannian manifolds.* Math. Z. **263** (2009), 33–67.

[15] O. Druet, E. Hebey, F. Robert, *Blow-up theory for elliptic PDEs in Riemannian geometry.* Mathematical Notes **45**, Princeton University Press, 2004.

[16] O. Druet, E. Hebey, J. Vétois, *Bounded stability for strongly coupled critical elliptic systems below the geometric threshold of the conformal Laplacian.* J. Funct. Anal. **258** (2010), 999–1059.

[17] P. Esposito, A. Pistoia, J. Vétois, *The effect of linear perturbations on the Yamabe problem.* Math. Ann. (2013), doi: 10.1007/s00208-013-0971-9.

[18] Z.-C. Han, *Asymptotic approach to singular solutions for nonlinear elliptic equations involving critical Sobolev exponent.* Ann. Inst. H. Poincaré Anal. Non Linéaire **8** (1991), 159–174.

[19] E. Hebey, M. Vaugon, *From best constants to critical functions.* Math. Z. **237** (2001), 737–767.

[20] E. Hebey, J. Wei, *Resonant states for the static Klein–Gordon–Maxwell–Proca system.* Math. Res. Lett. **19** (2012), 953–967.

[21] M.A. Khuri, F.C. Marques, R.M. Schoen, *A compactness theorem for the Yamabe problem.* J. Differential Geom. **81** (2009), 143–196.

[22] J.M. Lee, T.H. Parker, *The Yamabe problem.* Bull. Amer. Math. Soc. (N.S.) **17** (1987), 37–91.

[23] Y.-Y. Li, *On a singularly perturbed elliptic equation.* Adv. Differential Equations **2** (1997), 955–980.

[24] Y.-Y. Li, L. Zhang, *A Harnack type inequality for the Yamabe equation in low dimensions.* Calc. Var. Partial Differential Equations **20** (2004), 133–151.

[25] Y.-Y. Li, L. Zhang, *Compactness of solutions to the Yamabe problem. II.* Calc. Var. Partial Differential Equations **24** (2005), 185–237.

[26] Y.-Y. Li, L. Zhang, *Compactness of solutions to the Yamabe problem. III.* J. Funct. Anal. **245** (2007), 438–474.

[27] Y.-Y. Li, M.-J. Zhu, *Yamabe type equations on three-dimensional Riemannian manifolds.* Commun. Contemp. Math. **1** (1999), 1–50.

[28] F.C. Marques, *A priori estimates for the Yamabe problem in the non-locally conformally flat case.* J. Differential Geom. **71** (2005), 315–346.

[29] A.M. Micheletti, A. Pistoia, J. Vétois, *Blow-up solutions for asymptotically critical elliptic equations.* Indiana Univ. Math. J. **58** (2009), 1719–1746.

[30] M. Musso, A. Pistoia, *Multispike solutions for a nonlinear elliptic problem involving the critical Sobolev exponent.* Indiana Univ. Math. J. **51** (2002), 541–579.

[31] A. Pistoia, G. Vaira, *On the Stability for the Paneitz–Branson equation.* Int. Math. Res. Notices (2012), doi: 10.1093/imrn/rns133.

[32] A. Pistoia, J. Vétois, *Sign-changing bubble towers for asymptotically critical elliptic equations on Riemannian manifolds.* J. Differential Equations **254** (2013), 4245–4278.

[33] S.I. Pohožaev, *On the eigenfunctions of the equation* $\Delta u + \lambda f(u) = 0$. Soviet. Math. Dokl. **6** (1965), 1408–1411.

[34] O. Rey, *Proof of two conjectures of H. Brézis and L.A. Peletier.* Manuscripta Math. **65** (1989), 19–37.

[35] O. Rey, *The role of the Green's function in a nonlinear elliptic equation involving the critical Sobolev exponent.* J. Funct. Anal. **89** (1990), 1–52.

[36] F. Robert, J. Vétois, *Sign-changing blow-up for scalar curvature type equations.* Comm. Partial Differential Equations **38** (2013), 1437–1465.

[37] R.M. Schoen, *Conformal deformation of a Riemannian metric to constant scalar curvature.* J. Differential Geom. **20** (1984), 479–495.

[38] R.M. Schoen, *Variational theory for the total scalar curvature functional for Riemannian metrics and related topics.* Topics in calculus of variations, Montecatini Terme, 1987. Lecture Notes in Math. **1365**, 120–154, Springer, 1989.

[39] R.M. Schoen, S.-T. Yau, *On the proof of the positive mass conjecture in general relativity.* Comm. Math. Phys. **65** (1979), 45–76.

[40] R.M. Schoen, S.-T. Yau, *Conformally flat manifolds, Kleinian groups and scalar curvature.* Invent. Math. **92** (1988), 47–71.

[41] N.S. Trudinger, *Remarks concerning the conformal deformation of Riemannian structures on compact manifolds.* Ann. Scuola Norm. Sup. Pisa (3) **22** (1968), 265–274.

[42] H. Yamabe, *On a deformation of Riemannian structures on compact manifolds.* Osaka Math. J. **12** (1960), 21–37.

Pierpaolo Esposito
Dipartimento di Matematica
Università degli Studi "Roma Tre"
Largo S. Leonardo Murialdo 1
I-00146 Rome, Italy
e-mail: esposito@mat.uniroma3.it

Angela Pistoia
Dipartimento SBAI
Università di Roma "La Sapienza"
Via Antonio Scarpa 16
I-00161 Rome, Italy
e-mail: pistoia@dmmm.uniroma1.it

Jérôme Vétois
Laboratoire J.-A. Dieudonné
UMR CNRS-UNS N 7351
Université de Nice – Sophia Antipolis
Parc Valrose
F-06108 Nice Cedex 2, France
e-mail: vetois@unice.fr

Concentration Analysis and Applications to PDE
Trends in Mathematics, 49–67

Extremals for Sobolev and Exponential Inequalities in Hyperbolic Space

Gianni Mancini and Kunnath Sandeep

To Antonio

Abstract. We review results concerning optimal Sobolev inequalities in Riemannian manifolds and recent existence/non existence/uniqueness results for Sobolev extremals in the hyperbolic space $\mathbb{H}^n$. We also discuss exponential integrability in $\mathbb{H}$, the hyperbolic plane, and related topics.

Mathematics Subject Classification (2010). 35J20 (35A23 35J91 42B37).

Keywords. Sharp Sobolev and Moser inequalities; hyperbolic space.

1. Introduction

We start reviewing known results concerning Sobolev inequalities in Riemannian manifolds (M, g) of dimension n. If $n \geq 3$ and $2^* = \frac{2n}{n-2}$, embedding of $H^1(M)$ into $L^{2^*}(M)$ holds true on any *compact* (M, g): $\exists A, B \in \mathbb{R}$ such that

$$\|u\|^2_{L^{2^*}(M)} \leq A \int_M |\nabla_g u|^2 dv_g + B \int_M u^2 dv_g \qquad \forall u \in H^1(M) \tag{1.1}$$

(see [27]). However, if M is non compact, $H^1(M)$ is not continuously embedded in $L^{2^*}(M)$, in general. If M is complete, a necessary condition, found by Carron [15], is that

$$\inf_{x \in M} \mathrm{Vol}\,(B_1(x)) > 0. \tag{$*$}$$

Condition $(*)$ is never satisfied if M is *non compact* and has *finite volume*: see [27, Prop. 3.5]; but $(*)$ might fail as well in manifolds of infinite volume.

In turn, property $(*)$ becomes sufficient if, in addition, the Ricci curvature is bounded from below ([27, Thm. 3.2]). However, a bound on the curvature is not necessary. A sufficient, and also necessary condition, to have a Sobolev embedding

This work was partially supported by Prin "Problemi con piccoli divisori e perdita di compattezza in sistemi hamiltoniani ed equazioni alle derivate parziali nonlineari".

in a *complete manifold with infinite volume M* is the validity of the Faber–Krahn inequality [15]:

$$\inf_{\Omega\subset\subset M} \lambda_1(\Omega)\,\mathrm{Vol}_g(\Omega)^{\frac{2}{n}} > 0$$

where $\lambda_1(\Omega)$ is the first eigenvalue of Δ_g in Ω under Dirichlet boundary conditions.

Assuming (1.1), we look for optimal constants. Let $\mathcal{P}:=\{(A,B):(1.1) \text{ holds}\}$. It is known, since the work of Aubin [6], that, if M is compact or M is complete with *constant sectional curvature and positive injectivity radius*, then

$$A_0 := \frac{4}{n(n-2)\omega_n^{\frac{2}{n}}} = \min\{A : (A,B) \in \mathcal{P}\} \tag{1.2}$$

(ω_n :=volume of S^n). Carrying over the Aubin program, Hebey proved, among other things, that $A_0 = \inf\{A : (A,B) \in \mathcal{P}\}$ if M is *complete, with Ricci curvature bounded from below and positive injectivity radius* ([26]) and it is achieved if, in addition, the *Ricci curvature is bounded and g is conformally flat outside a compact set* [28].

Thus, A_0 is an optimal constant for a large class of manifolds. For these manifolds, if $B_0(M) := \inf\{B : (A_0(M), B) \in \mathcal{P}\}$, it results

$$B_0(M) \geq \frac{\max_M \mathrm{Scal}_{(M,g)}}{n(n-1)\omega_n^{\frac{2}{n}}} \quad \text{if } n \geq 4$$

$$B_0(M) \leq \frac{\max_M \mathrm{Scal}_{(M,g)}}{n(n-1)\omega_n^{\frac{2}{n}}} \quad \text{if } M \text{ can be conformally immersed into } S^n$$

(see [27, Prop. 5.1] and [46], respectively). When B_0 is achieved, inequality (1.1) becomes *totally optimal*:

A_0 *cannot be improved* and $\exists B < B_0 : (A,B) \in \mathcal{P} \Rightarrow A > A_0$

and it becomes relevant to identify extremal functions for

$$\|u\|^2_{L^{2^*}(M)} \leq A_0(M)\int_M |\nabla_g u|^2 dv_g + B_0(M)\int_M u^2 dv_g. \tag{OPT}$$

We will present here, in the first part, some results in case $(M,g) = (\mathbb{H}^n, h)$, the n-dimensional hyperbolic space, $n \geq 3$, for which clearly (1.1) holds true.

In the limiting case $n = 2$, Sobolev inequalities lead, in compact manifolds, to inequalities of exponential type (say, Moser–Trudinger type inequalities)

$$\exists C, \mu > 0 : \qquad \int_M e^u dv_g \leq C e^{\mu \int_M (|\nabla u|^2 + u^2) dv_g} \qquad \forall u \in H^1(M).$$

We will briefly recall sharp Moser–Trudinger inequalities in the compact setting and then we will move to Moser inequalities in $(\mathbb{H}^n, h)$, and related topics.

2. Sobolev inequalities in $\mathbb{H}^n, n \geq 3$

In $\mathbb{H}^n, n \geq 4$, the totally optimal Sobolev inequality (OPT) reads as

$$S\|u\|^2_{L^{2^*}(\mathbb{H}^n)} \leq \int_{\mathbb{H}^n} \left[|\nabla u|^2 - \frac{n(n-2)}{4} u^2 \right] dv_h \qquad \forall u \in H^1(\mathbb{H}^n) \tag{2.3}$$

where$S = A_0^{-1}$ is the best Sobolev constant in the Euclidean Sobolev inequality and $-\frac{n(n-2)}{4} = SB_0 = \frac{(n-2)}{4(n-1)} \omega_n^{-\frac{2}{n}} \max_M \mathrm{Scal}_{(M,g)}$.

Notice that (2.3) implies a Poincaré inequality; however, the best constant in the Poincaré inequality in $\mathbb{H}^n$ is $(\frac{n-1}{2})^2$ $\left(\text{and not } \frac{n(n-2)}{4}\right)$:

$$\int_{\mathbb{H}^n} |\nabla u|^2 dv_h \geq \left(\frac{n-1}{2}\right)^2 \int_{\mathbb{H}^n} u^2 dv_h \qquad \forall u \in H^1(\mathbb{H}^n). \tag{2.4}$$

As a consequence, one can improve the second constant but at expenses of the first constant: $\forall\, \lambda \in (\frac{n(n-2)}{4}, \frac{(n-1)^2}{4})$, $\exists\, S_\lambda \in (0, S]$ such that

$$S_\lambda \|u\|^2_{L^{2^*}(\mathbb{H}^n)} \leq \int_{\mathbb{H}^n} \left[|\nabla u|^2 - \lambda u^2 \right] dv_h \qquad \forall u \in H^1(\mathbb{H}^n). \tag{2.5}$$

A simple derivation of (2.4) and (2.5) from Hardy and Hardy–Sobolev–Maz'ja inequalities (see [37], [16]), can be found in [34].

If $n \geq 4$ and $\lambda \in (\frac{n(n-2)}{4}, \frac{(n-1)^2}{4})$, necessarily $S_\lambda < S$, because (2.3) is totally optimal (i.e., S and $\frac{n(n-2)}{4}$ are sharp constants). However, if $n = 3$, then $S_\lambda = S$ for every $\lambda \leq 1$. This has been shown in [12] (see also [34]).

Going back to (2.5), it can be shown that also $\lambda = (\frac{n-1}{2})^2$ is allowed, and then it is, clearly, the optimal value for the "second constant" in Sobolev inequalities in $\mathbb{H}^n$ (this also implies that Δ_h – *the hyperbolic Laplacian, i.e., the Laplace–Beltrami operator on* $\mathbb{H}^n$ – has no L^2 eigenfunction corresponding to the "first eigenvalue" $\lambda_1(\mathbb{H}^n) = (\frac{n-1}{2})^2$).

As for extremals, the optimal inequality (2.3) does not possess any.

Conversely, the other limiting inequality (corresponding to $\lambda = (\frac{n-1}{2})^2$) has, if $n \geq 4$, a unique extremal function, which however is not in $H^1(\mathbb{H}^n)$: it belongs to $\mathcal{H}$, the closure of $C_0^\infty(\mathbb{H}^n)$ w.r.t. $|||u|||^2 := \int_{\mathbb{H}^n} \left[|\nabla u|^2 - (\frac{n-1}{2})^2 u^2 \right] dv_h$ (it is a norm thanks to (2.5)). We state these results in the next subsection.

2.1. Symmetry and uniqueness of extremals for (2.5) [34]

We first recall a symmetry and uniqueness result for extremals of (2.5), and in fact, more in general, for entire positive solutions of the equations

$$\Delta_h u + \lambda u + |u|^{p-2} u = 0 \tag{2.6}$$

where $n \geq 3$ and $p \in (2, 2^*]$ (see [34] for the case $n = 2$).

Hyperbolic symmetry ([16] and [34]). *Let* $\lambda \le \frac{(n-1)^2}{4}$. *Let v be a positive entire solution of* (2.6). *Then v has hyperbolic symmetry, i.e.,*

$$\exists x_0 \in \mathbb{H}^n \text{ such that } v(x) \text{ depends only on the distance between } x_0, x \in \mathbb{H}^n.$$

Here, u is a positive entire solution meaning the following: if $\lambda < (\frac{n-1}{2})^2$, then $u \in H^1(\mathbb{H}^n)$, if $\lambda = (\frac{n-1}{2})^2$, then $u \in \mathcal{H}$, the space introduced above.

By hyperbolic symmetry, (2.6) rewrites, in hyperbolic polar coordinates, as

$$w'' + (n-1)\coth t\, w' + \lambda w + w^{p-1} = 0. \tag{2.7}$$

Now, an application of Kwong's "energy method" [16] to (2.7) gives a complete picture to uniqueness problems for (2.6). The crucial information is given by the asymptotic behaviour of positive entire solutions:

$$\lim_{t\to+\infty} \frac{\log w^2}{t} = \lim_{t\to+\infty} \frac{\log w'^2}{t} = -\left[n-1+\sqrt{(n-1)^2-4\lambda}\right].$$

Uniqueness. (2.6) *has at most one positive entire solution.*

2.2. Existence/nonexistence of extremals for (2.5) [34]

Some phase analysis and an application of the maximum principle give

A general non existence result. *Let* $\lambda \le \frac{n(n-2)}{4}$ *and* $p = 2^*$ *or* $\lambda > (\frac{n-1}{2})^2$. *Then* (2.6) *has no positive entire solution.*

Existence for $n > 3$ ([47], [34]). (2.6) *has a positive entire solution iff* $\lambda \le \frac{(n-1)^2}{4}$ *in case* $p < 2^*$ *and iff* $\lambda \in (\frac{n(n-2)}{4}, \frac{(n-1)^2}{4}]$ *in case* $p = 2^*$.

Positive entire solutions can be searched as minimizer for $S_\lambda = \inf_N J$, where

$$J(u) = \frac{\int_{\mathbb{H}}[|\nabla u|^2 - \lambda u^2]dv_h}{\left(\int_{\mathbb{H}} |u|^{2^*} dv_h\right)^{\frac{2}{2^a st}}}, \qquad N = \{u \neq 0; \int_{\mathbb{H}} [|\nabla u|^2 - \lambda u^2]dv_h = \int_{\mathbb{H}} |u|^{2^*} dv_h\}$$

if $p = 2^*$. Assuming S_λ is not achieved, a concentration-compactness analisys leads to $S_\lambda \ge S$, a fact which, as observed above, is not possible if $n \ge 4$ and $\lambda \in (\frac{n(n-2)}{4}, \frac{(n-1)^2}{4}]$ (as for the subcritical case, see [34] and [21]).

A striking low-dimension phenomena [34]. *Let* $n = 3$. *Then* (2.6) *has no positive entire solution for any* λ. *In particular,* $S_\lambda = S$ *for every* $\lambda \le 1$.

See also [12] for nonexistence of minimizers. Notice that this result implies that, in Proposition 5.1 in [27], assumption $n \ge 4$ cannot be dropped.

3. The limiting case $n = 2$: Sobolev inequalities and exponential integrability in the compact setting

If (M, g) is a compact 2-manifold, then H^1 is continuously embedded in L^p for every p and sharp Sobolev inequalities lead to (sharp) exponential inequalities. A striking example is given by Beckner's derivation of the celebrated Moser–Onofri inequality.

3.1. The Moser–Onofri inequality [39], [41]

$$J(u) := \frac{1}{|S^2|}\int_{S^2}\left[2u + |\nabla u|^2\right]d\sigma - \ln\left(\frac{1}{|S^2|}\int_{S^2} e^{2u}d\sigma\right) \geq 0 \qquad \text{(MO)}$$

$\forall u \in H^1(S^2)$.

Beckner [9] derived (MO) from the sharp Sobolev inequalities

$$\left(\int_{S^2}|w|^p d\sigma\right)^{\frac{2}{p}} \leq \frac{p-2}{2}\int_{S^2}|\nabla w|^2 + \int_{S^2} w^2 d\sigma \qquad \forall p \geq 2$$

where σ is the normalized measure satisfying $\int_{S^2} d\sigma = 1$. He takes $w_p = 1 + \frac{u}{p}$ and sends p to infinity to get (MO).

Another, more recent, proof of the Moser–Onofri inequality has been given by Dolbeault [22] starting from the *sharp Gagliardo–Nirenberg inequalities in* $\mathbb{R}^2$

$$\|w\|_{2p}^{2p} \leq C_p\|\nabla w\|_2^{p-1}\,\|w\|^{p+1} \qquad \forall w \in H^1(\mathbb{R}^2). \qquad \text{(GN)}$$

He takes $w_p = F_p\left(1 + \frac{u}{2p}\right)$ where $F_p = (1+|x|^2)^{\frac{1}{p-1}}$ gives equality in (GN). Since $F_p^{2p} \to \frac{1}{(1+|x|^2)^2}$, he finds in the limit

$$\log\left(\int_{\mathbb{R}^2}\frac{e^u}{\pi(1+|x|^2)^2}dx\right) \leq \int_{\mathbb{R}^2}\frac{u}{\pi(1+|x|^2)^2}dx + \frac{1}{16\pi}\int_{\mathbb{R}^2}|\nabla u|^2 dx$$

which is (MO) in stereographic coordinates.

We recall now a few facts on (MO) and the importance to find related extremals.

Moser introduced (MO) to try to identify functions K on S^2 which are the Gauss curvature of some conformal metric $g = e^{2u}g_0$ (see [38]). In fact, this amounts to find critical points u of

$$J_K(u) := \frac{1}{|S^2|}\int_{S^2}\left[2u + |\nabla u|^2\right]d\sigma - \ln\left(\frac{1}{|S^2|}\int_{S^2} Ke^{2u}d\sigma\right).$$

Easily, (MO) implies $\inf J_K > -\infty$. However J_K has no global minimum (see [18]). Still, (MO) was a crucial step in the search of critical points of J_K.

If $K \equiv 1$, extremals for (MO) are the solutions of the geometric equation

$$-\frac{1}{8\pi}\Delta u + \mu = \lambda\mu e^u = e^{u+\log\mu+\log\lambda}$$

where $\mu = \frac{1}{\pi(1+|x|^2)^2}$ and $\lambda = \int_{\mathbb{R}^2}\mu e^u$. Since$-\Delta(\log\mu) = 8\pi\mu$,$u$ is an extremal iff $v = u + \log\mu$ is a solution of the Liouville equation

$$-\Delta v = ce^v$$

for some $c > 0$. Notice that $e^v = \mu e^u$ is in L^1 and such solutions have been identified by Chen and Li [20], and v is "essentially unique". So, extremals are completely classified, and are given by

$$u = \frac{1}{2}\log|\det d\phi|, \qquad \phi \text{ any conformal diffeomorphism on } S^2.$$

Another geometric consequence of (MO) stems from the Polyakov–Ray–Singer formula

$$\log\left(\frac{\det(-\Delta_{g_0})}{\det(-\Delta_{g_u})}\right) = \frac{1}{3}J(u). \tag{PRS}$$

Recall that, if λ_k are the positive eigenvalues of $-\Delta_g$,

$$\log\det(-\Delta_g) := \log\prod_{k=1}^{\infty}\lambda_k = -\zeta'(0)$$

(at least formally) where ζ is the "regularized" zeta function

$$\zeta(s) = \sum_{\lambda_k\neq 0}\frac{1}{\lambda_k^s} = \frac{1}{\Gamma(s)}\int_0^{\infty}\sum_{k=1}^{\infty}e^{-\lambda_k t}\,t^{s-1}dt.$$

The Moser–Onofri inequality implies, jointly with the Polyakov–Ray–Singer formula, that g_0, the Euclidean metric on S^2, is a maximizer for $\det(-\Delta_g)$ in the class of all conformal metrics (see [17], [19]).

3.2. The Moser–Trudinger inequality

Moser [39] derived (MO) – in fact a weaker form of it – as a "linearization" of

$$M(S^2) := \sup_{\int_{S^2}|\nabla u|^2\le 1,\ \int_{S^2}u=0} M(u,S^2) < +\infty$$

where

$$M(u,S^2) := \int_{S^2}\left[e^{4\pi u^2}-1\right]d\sigma.$$

This is a sharp version (4π cannot be improved!) of the *Trudinger inequality* [48]. We will refer to $M(u) = M(u,S^2)$ as the *M-functional* (on S^2).

Moser also gave a version in bounded domains $\Omega\subset\mathbb{R}^2$:

$$M(\Omega) := \sup_{\int_\Omega|\nabla u|^2\le 1,\ u\in H_0^1(\Omega)}\int_\Omega[e^{4\pi u^2}-1]dx < +\infty \tag{MT}$$

and then derived a "linearized" version:

$$\exists A\ge 0:\qquad I(u) := \frac{1}{16\pi}\int_\Omega|\nabla u|^2 - \log\left[\frac{1}{|\Omega|}\int_\Omega e^u dx\right] + A \ge 0 \tag{MTL}$$

$\forall u\in H_0^1(\Omega)$. In (MTL), $\frac{1}{16\pi}$ is a sharp constant, i.e., it cannot be improved.

In case Ω is a ball, **the best constant** A is equal to 1 (see [14]) and *it is not achieved!* In fact, if (MTL) has a minimizer v, then, by symmetrization, we can find *a positive radial minimizer u* for (MTL) on a ball B_R. It has to satisfy the Euler–Lagrange equation

$$-\Delta u = \lambda\frac{e^u}{\int_B e^u},\qquad \text{(with } \lambda = 8\pi).$$

However such an equation has only the positive radial solution given by

$$u_\lambda(r) = \log\left(\frac{8\pi}{8\pi - \lambda + \lambda\left(\frac{r}{R}\right)^2}\right)^2$$

and *such a solution exists iff* $\lambda < 8\pi$: inequality (MTL) is always strict [29].

An extension of (MT) involving higher derivatives is due to Adams [2].

An extension to *n-dimensional compact* Riemannian manifolds (X, g) (with higher derivatives) was given by Fontana [25]:

$$M(X) := \sup_{\int_X |\nabla_g u|^n \leq 1,\ \int_X u=0} \int_X \left[\exp\left(n\omega_{n-1}^{\frac{1}{n-1}} u^{\frac{n}{n-1}}\right) - 1\right] dv_g < +\infty. \quad \text{(MTX)}$$

Here, ω_n denotes the area of S^n.

3.3. Extremal funtions for the M-Functional $M(u, X) := \int_X [e^{4\pi u^2} - 1] dv_g$

To begin with, let $X = D$ be the unite disc in $\mathbb{R}^2$. Let us introduce the *Moser functions*

$$\mathcal{M}(r) := (2\pi)^{-\frac{1}{2}}\left[\chi_{[0,\frac{1}{e})} - \chi_{[\frac{1}{e},1]}\log r\right], \quad \mathcal{M}_\epsilon(r) := \epsilon^{-\frac{1}{2}}\mathcal{M}(r^\epsilon).$$

We have $\|\nabla\mathcal{M}_\epsilon\|_2 = 1, \|\mathcal{M}_\epsilon\|_2 = \circ(1)$while $M(\mathcal{M}_\epsilon) = 3\pi + \circ(1)$, so, $M(u, D)$ *is not weakly continuous* (according to [4] this is essentially the only way $M(u, D)$ loses compactness): this makes it difficult to find extremals.

The first existence result is due to Carleson and Chang [14]. It is based on

P.L. Lions "Converge or concentrate" Lemma (on a bounded domain Ω) [32]. *Let $\Omega \subset \mathbb{R}^2$ be a bounded domain. Let $u_k \rightharpoonup u$ in $H_0^1(\Omega)$, $\|\nabla u_n\|_2 \leq 1$. Then,*

$$\text{either} \quad u \neq 0 \quad \text{and} \quad M(u_k, \Omega) \to M(u, \Omega)$$
$$\text{or} \quad u = 0 \quad \text{and} \quad \exists x_0 \in \overline{\Omega} : |\nabla u_k|^2 \rightharpoonup \delta_{x_0}.$$

After defining $Cl(D) :=$ *the highest concentration level* for $M(u, D)$, i.e.,

$$u_k \rightharpoonup u, \quad M(u_k, D) \to c > Cl(D) \quad \Rightarrow \quad M(u_k, D) \to M(u, D).$$

Carleson and Chang were able to prove that $Cl(D) = \pi e$.

Finally, they built a test function u_M s. t. $M(u_M, D) > \pi e$, to conclude:

$$M(D) := \sup\{M(u, D) : \quad \|\nabla u\|_2 \leq 1\} \text{ is achieved.}$$

Flucher [24] used the Carleson–Chang result to find extremals for any (bounded) domain Ω.

In case $\Omega = \varphi(D)$,φ a conformal diffeomorphism, i.e., Ω is *simply connected*, let $cr_\Omega(\varphi(z)) := (1 - |z|^2)|\varphi'(z)|$ be the *conformal radius* at $x = \varphi(z)$.

Set $cR(\Omega) := \sup_{x\in\Omega} cr_\Omega(x)$. Flucher found that

i) $Cl(\Omega) = cR^2(\Omega)\, Cl(D)$, ii) $M(\Omega) \geq cR^2(\Omega)\, M(D)$.

One can obtain i) by a change of variables while ii) follows by subharmonicity of $|\varphi'|^2$. So, $M(\Omega) > Cl(\Omega)$ and $M(\Omega)$ is achieved by Lions' Lemma.

For general Ω, Flucher uses an extension of $cr_\Omega(x)$, the *harmonic radius*

$hr_\Omega(x) = e^{-2\pi H_{\Omega,x}(x)}$, $H_{\Omega,x}$ the regular part of the Green function.

A proof for $M(u, X)$ on a compact Riemannian manifold X can be found in [31]. The main claim, based on a blow-up analysis, says that

$$Cl(X) = \pi e^{1-4\pi H_{X,x}(x)}, \quad \text{where} \quad H_{X,x}(y) = -\frac{1}{2\pi}\log \operatorname{dist}(x,y) - G_x(y)$$

and

$$-\Delta G_x(y) = \delta_x - \frac{1}{\operatorname{vol} X}, \qquad \int_X G_x = 0, \qquad \inf G_x > -\infty.$$

4. The limiting case $n = 2$: Exponential integrability in manifolds of infinite measure

Exponential inequalities clearly imply the Poincaré inequality:

$$\exists \alpha > 0: \quad C := \sup \int_X [e^{\alpha(\frac{u}{\|\nabla u\|_2})^2} - 1] < \infty \quad \Rightarrow \quad \int_X |\nabla u|^2 \geq \frac{\alpha}{C}\int_X |u|^2.$$

But the converse is not always true, and in fact little is known on sharp Moser–Trudinger inequalities on manifolds of infinite measure, even in the simple case of open subsets of $\mathbb{R}^2$.

4.1. Scale invariant M-T inequality in $\mathbb{R}^2$ and Orlicz embeddings

A (subcritical) scale invariant Moser-type inequality (in the whole of $\mathbb{R}^2$) was given by Adachi–Tanaka [1], and a (critical) H^1 version is due to Ruf [45]:

$$\int_{\mathbb{R}^2} |\nabla u|^2 \leq 1 \Rightarrow \int_{\mathbb{R}^2} \left(e^{\alpha u^2} - 1\right) dx \leq c_\alpha \int_{\mathbb{R}^2} |u|^2 dx \quad \text{if} \quad \alpha < 4\pi, \tag{AT}$$

$$M_1(\mathbb{R}^2) := \sup\{\int_{\mathbb{R}^2} \left(e^{4\pi u^2} - 1\right) dx : \quad \|u\|_{H^1} \leq 1\} < \infty. \tag{R}$$

4π is optimal in both inequalities, but it is not allowed in (AT).

By symmetrization, it is enough to prove (AT) (or (R)) for radially symmetric decreasing functions: they enjoy the asymptotic decay $u(|x|) \leq \frac{\|u\|_2}{|x|}$ which, in turn, gives a uniform control on $\int_{|x|\geq R}(e^{4\pi u^2} - 1)$.

Both (AT) and (R) readily imply the 2D critical Sobolev embedding in the Orlicz space $\mathcal{L}$. Let $\int_{\mathbb{R}^2}(e^{\frac{u^2}{\lambda(u)}} - 1)dx = M(\mathbb{R}^2)$. Then $\|u\|_{\mathcal{L}} := \lambda(u)$, and

$$\|u\|_{\mathcal{L}} \leq \frac{1}{\sqrt{4\pi}}\|u\|_{H^1} \qquad (1/\sqrt{4\pi} \quad \text{is a sharp constant}). \tag{$\mathcal{L}$}$$

If $\mathcal{M}_\epsilon$ is as in 3.3, $\|\mathcal{M}_\epsilon\|_{\mathcal{L}} \to \frac{1}{\sqrt{4\pi}}$ and hence this embedding is not compact. More in general, lack of compactness (in $H^1_{0,\mathrm{rad}}(D)$) can be obtained via the action

$$\epsilon \to u_\epsilon := \frac{1}{\sqrt{\epsilon}}u(r^\epsilon).$$

Notice that the "bubbles" u_ϵ have the properties

$$\int_D |\nabla u_\epsilon|^2 = \int_D |\nabla u|^2, \quad u_\epsilon \rightharpoonup 0 \qquad \text{and} \qquad \|u_\epsilon\|_{\mathcal{L}} \geq \delta > 0.$$

Actually, Bahouri, Majdoub and Masmoudi [7] have described how, in some special case, compactness fails. Assuming

$$\lim_{R\to\infty}\left[\limsup_{n\to\infty}\int_{|x|\geq R}|w_n|^2\right] = 0,$$

they show that *if* $w_n \in H^1_{0,\mathrm{rad}}(\mathbb{R}^2)$, $w_n \rightharpoonup 0$ *but* $\limsup_n \|w_n\|_{\mathcal{L}} > 0$, *then* w_n *are (up to a small error in* $\mathcal{L}$*-norm) superpositions of 'bubbles'* u_ϵ.

Remark 4.1. A function u is an extremal for ($\mathcal{L}$) iff it is an extremal for the Moser functional $M(u,\mathbb{R}^2)$ on $\{\|u\|_{H^1} \leq 1\}$: actually, *it is not known wether such extremal functions exist.*

4.2. Sobolev and Moser inequalities in the hyperbolic plane $\mathbb{H}$

In 1998, in a paper concerning weighted Sobolev inequalites arising in fluid mechanics, Talenti and Mugelli [40] computed, in some cases, sharp positive constants $S(p,\lambda)$ for related Sobolev inequalities in the hyperbolic plane:

$$S(p,\lambda)\left(\int_{\mathbb{H}}|u|^p dv_h\right)^{\frac{2}{p}} \leq \int_{\mathbb{H}}\left[|\nabla_{\mathbb{H}} u|^2 - \lambda u^2\right] dv_h \qquad \forall u \in C_0^\infty(\mathbb{H}).$$

Here, $p > 2$, $\lambda \leq \frac{1}{4}$. They find, e.g., (see also Beckner [9]), that

$$\frac{\pi^{\frac{2}{3}}}{4}\|u\|^2_{L^6(\mathbb{H})} \leq \int_{\mathbb{H}}\left[|\nabla u|^2 - \frac{3}{16}u^2\right] dv_h \qquad \forall\, u \in C_0^\infty(\mathbb{H}) \tag{4.8}$$

$$\frac{3\pi^{\frac{2}{3}}}{4}\|u\|^2_{L^6(\mathbb{H})} \leq \int_{\mathbb{H}}\left[|\nabla u|^2 - \frac{1}{4}u^2\right] dv_h \qquad \forall\, u \in C_0^\infty(\mathbb{H}) \tag{4.9}$$

and identify extremals, which, however, are not in $H^1(\mathbb{H})$ (e.g., (4.9) has an extremal in $\mathcal{H}$, the function space introduced in Section 3): according to Beckner "the right-hand sides in (4.8)–(4.9) are to be evaluated as limiting forms for functions that may not be in $L^2(\mathbb{H})$". In [40] the AA's also prove hyperbolic symmetry and, in some cases, uniqueness, of extremals.

In [34], we extended the symmetry result for the extremals up to the limiting value $\lambda = \frac{1}{4}$ and the uniqueness results up to $\lambda \leq \frac{2p}{(p+2)^2}$.

Finally, Talenti–Mugelli obtained the estimate, for $S_p := S(p,0)$,

$$S_p \geq \left(\frac{8\pi p\Gamma^2(\frac{p}{p-2})}{(p-2)^2\Gamma(\frac{2p}{p-2})}\right)^{1-\frac{2}{p}} = \frac{8\pi}{p}(1+\circ(1)) \quad \text{as} \quad p \to +\infty$$

which gives a Trudinger inequality below $\frac{4\pi}{e}$.

In [35] (see also the work by Adimurti–Tintarev [3]) we proved the

Moser–Trudinger inequality in $\mathbb{H}$:

$$M(\mathbb{H}) := \sup\left\{\int_{\mathbb{H}}\left[e^{4\pi u^2}-1\right]dv_h : \int_{\mathbb{H}}|\nabla_h u|^2 dv_h \le 1\right\} < \infty.$$

An (AT) type inequality holds as well: $\forall\alpha < 4\pi$, it exists $c_\alpha > 0$ s. t.

$$\int_{\mathbb{H}}|\nabla u|^2 dv_h \le 1 \quad\Rightarrow\quad \int_{\mathbb{H}}\left(e^{\alpha u^2}-1\right)dv_h \le c_\alpha\int_{\mathbb{H}}u^2 dv_h \tag{4.10}$$

$\forall u \in H_0^1(\mathbb{H})$.

To get these inequalities, we reduce estimates to functions in $H^1_{0,\text{rad}}(D)$ via *symmetrization with respect to the hyperbolic measure* v_h,

$$u_h^*(x) := \sup\{t \ge 0 : v_h(\{u > t\}) > v_h(\{d_h(x,0)\}\}$$

(d_h =hyperbolic distance).

For $u \in H^1_{0,\text{rad}}(D)$, one gets easily $2\pi u^2(r) \le \log\frac{1}{r}\int_D|\nabla u|^2dx$. Then,

$$\int_{|x|\ge r}\left[e^{4\pi u^2}-1\right]dv_h \le 16\pi\int_{|x|\ge r}u^2 dv_h \forall r \in \left(\frac{1}{2},1\right).$$

This, jointly with the Euclidean (AT), gives (4.10). On the other hand,

$$\int_{\mathbb{H}}|\nabla_h u|^2 \ge \frac{1}{4}\int_{\mathbb{H}}|u|^2 dv_h \qquad \textit{Poincaré (Hardy) inequality}$$

and the Euclidean (MT) yields (MT) for $\mathbb{H}$.

More in general, we may consider *conformal metrics* $g = \rho h$ on $\mathbb{H}$, where $0 < \rho \in C^2$, with corresponding measure $dv_g = \rho dv_h$, *i.e., conformal discs* (D,g). Recall that the Dirichlet integral is a conformal invariant. We have

Theorem 4.2 (Moser inequality in conformal discs).

$$M(D,g) := \sup_{u\in C_0^\infty(D),\quad \int_D|\nabla u|^2dx\le 1}\int_D\left(e^{4\pi u^2}-1\right)dv_g < \infty$$

if and only if $\exists c > 0 : \ g \le c\,h$,*where* h *is the hyperbolic metric.*

Proof. ([35]) We have just to show that the bound $g \le ch$ is necessary. Working in the disc model, we assume by contradiction that

$$\inf_{|x_n|\le\epsilon}\rho(\varphi_n(x)) \ge \frac{1}{2}\rho(\varphi_n(0)) \to_n +\infty$$

for some conformal diffeomorphisms φ_n and some $\epsilon_n \to 0$. Taking a sequence of Moser functions $\mathcal{M}_n$ and because φ_n are hyperbolic isometries, we find

$$\int_D\left(e^{4\pi[\mathcal{M}_n\circ\varphi_n^{-1}]^2}-1\right)\zeta dv_h = \int_D\left(e^{4\pi\mathcal{M}_n^2}-1\right)\zeta\circ\varphi_n dv_h \to \infty. \qquad \square$$

Asymptotic estimate for best Sobolev constants in $\mathbb{H}$. The theorem above implies the following (sharp) improvment of the Mugelli–Talenti estimate:

$$S_p := \inf \frac{\int_{\mathbb{H}} |\nabla_h u|^2 dv_h}{\left(\int_{\mathbb{H}} |u|^p dv_h\right)^{\frac{2}{p}}} = \frac{8\pi e + \circ(1)}{p} \qquad (\text{as } p \to +\infty)$$

(see Ren and Wey [44] for the original result in domains of $\mathbb{R}^2$). Theorem 4.1 and the Taylor expansion yield $pS_p \geq \frac{8\pi e}{1+\circ(1)}$. The reverse inequality follows by testing S_p with Moser functions; same estimates hold true on any conformal disc if $g \leq Ch$.

Other consequences of Theorem 4.1 are the following

(LM) type inequalities in $\mathbb{H}$. $c = \frac{1}{16\pi}$ is the optimal constant in

$$\int_{\mathbb{H}} [e^u - u - 1] dv_h \leq \|\nabla u\|^2 \, e^{c\|\nabla u\|^2} \qquad \forall u \in H_0^1(\mathbb{H}).$$

Continuity of the Moser functional on $(\mathbb{H}, g)$, $g \leq ch$.

$$u \to u \text{ in } H_0^1(\mathbb{H}) \quad \Rightarrow \quad \int_{\mathbb{H}} \left(e^{4\pi u_n^2} - 1\right) dv_g \to \int_{\mathbb{H}} \left(e^{4\pi u^2} - 1\right) dv_g.$$

4.3. Moser inequality for $\Omega \subset \mathbb{R}^2$ of infinite measure

As already observed, if $\lambda_1(\Omega) = 0$ then $M(\Omega) := \sup_u M(u, \Omega) = \infty$. On the other hand, we see from (AT) that $\lambda_1(\Omega) > 0$ implies:

$$\sup_{u \in C_0^\infty(\Omega), \quad \int_\Omega |\nabla u|^2 \leq 1} \int_\Omega \left(e^{\alpha u^2} - 1\right) dx < \infty \qquad \forall \alpha < f 4\pi.$$

What about the critical case $\alpha = 4\pi$? We have a simple answer if Ω is *simply connected*, i.e., $\Omega = \varphi(D)$, φ a conformal diffeomorphism. By conformal invariance, $M(\Omega) = M(D, g)$, $g := |\varphi'|^2 g_e$ (g_e = Euclidean metric). Now, recall that $cr_\Omega(\varphi(z)) = |\varphi'(z)|\,(1 - |z|^2)$, while by Koebe's $\frac{1}{4}$-Theorem

$$\left[\textit{for every conformal diffeomorfism } \psi, \ \operatorname{dist}(\psi(0), \partial\Omega) \geq \tfrac{1}{4}|\psi'(0)|\right]$$

one finds that $cr_\Omega(x) \leq 4 \operatorname{dist}(x, \partial\Omega)$. Thus Theorem 4.1 yields

Corollary 4.3. *Let Ω be a* simply connected domain *in $\mathbb{R}^2$. Then*

$$iR(\Omega) < \infty \ \Leftrightarrow \ \lambda_1(\Omega) > 0 \ \Leftrightarrow \ M(\Omega) < \infty \ \Leftrightarrow \ cR(\Omega) < \infty$$

where $iR(\Omega) := \sup\{r > 0 : \exists D_r(x) \subset \Omega\} < \infty$ is the inner radius.

Remark 4.4. For bounded simply connected domains the first equivalence is well known (see [42], [43]). However it fails for general domains. A related question is the following: if Ω is any open subset of $\mathbb{R}^2$,

$$\lambda_1(\Omega) > 0 \quad \overset{?}{\Longleftrightarrow} \quad M(\Omega) < \infty \quad \overset{?}{\Longleftrightarrow} \quad hR(\Omega) := \sup_\Omega hr(x) < \infty.$$

Gabriele Mancini [33] answered in his master thesis the first question, showing that

Theorem 4.5. $\quad M(\Omega) < \infty \ \Leftrightarrow \ \lambda_1(\Omega) > 0 \qquad \forall \Omega \subset \mathbb{R}^2.$

Remark 4.6. In $(D, \rho g_h)$ the Poincaré inequality does not imply (MT).

The main tool in [33] is an improved decay estimate stemming from a local Polya–Szegő inequality:

$$v = u^* \Rightarrow \ v(R)^2 \leq \frac{1}{\pi R^2 \lambda_1(\Omega)} \left(1 - \int_{|x|\leq R} |\nabla v|^2\right).$$

5. Extremals for Moser inequalities in conformal discs

As already observed, maximizing sequences might not be compact. Still, due to Lions' Lemma, *the only obstruction in compact manifolds is concentration.* Carleson and Chang, first, and then Flucher, ruled out concentration in any bounded domain $\Omega \subset \mathbb{R}^2$ by showing that $M(\Omega) > Cl(\Omega)$, the highest concentration level. Flucher's arguments suggest that concentration shouldn't occur in unbounded domain either. However, compactness might fail for different reasons: *dichotomy or vanishing.*

Here is an example in $\mathbb{H} \cong (D, h)$: if

$$u_n = (1 - |x|^2)^{\frac{1}{2} - \frac{1}{n}}, \qquad w_n := \frac{u_n}{\|\nabla u_n\|_2}$$

then

$$\int_D |\nabla w_n|^2 dx = 1, \qquad \int_D w_n^2 = 4 + \circ(1) \quad w_n \rightharpoonup 0$$

uniformly and in H^1_{loc}, so w_n *does not concentrate: it vanishes* and

$$\lim_n \int_D (e^{4\pi w_n^2} - 1) dv_h = 16\pi.$$

Actually, the alternative is: converge or vanish.

We do not know if $M(\mathbb{H})$ is achieved, but we will show in Section 5.2 that $M(D_g)$ is achieved for some D_g uniformly conformally equivalent to the hyperbolic disc. But first we give quite a different example of a conformal disc where M is achieved.

5.1. The infinite strip and other conformal discs (from [11] and [33])

Let $\Sigma := (-r, r) \times \mathbb{R}$. Recall (see [8]) that

$$\lambda_1(\Sigma) = \frac{\pi^2}{4r^2}, \qquad cR^2(\Sigma) = \frac{16r^2}{\pi^2}.$$

Let

$$M(\Sigma, u_n) \to M(\Sigma) := \sup_{u \in C_0^\infty(\Sigma), \int_\Sigma |\nabla u|^2 \leq 1} \int_\Sigma \left(e^{4\pi u^2} - 1\right) dxdy$$

and $u_n \rightharpoonup u$. *Then* $u \neq 0$, *and* $M(\Sigma)$ *is achieved.*

An important role to get this result is played by the symmetry of Σ: up to a Steiner symmetrization (first in x and then in y) we can assume $u_n = u_n^*$, where $u = u^*$ means

(i) $0 \le u(-x, y) = u(x, y) = u(x, -y)$ and

(ii) $|x_1| \le |x_2|, |y_1| \le |y_2| \quad \Rightarrow \quad u(x_1, y_1) \le u(x_2, y_2)$.

In turn, symmetry implies the (crucial for the proof) asymptotic decay

$$u = u^* \quad \Rightarrow \quad u^2(x, y) \le \frac{c_r}{x} \|\nabla u\|_2^2. \tag{5.11}$$

The proof is carried over in three steps. First assume, by contradiction, $u = 0$.

Step 1. Concentration cannot occur. As a consequence $M(\Sigma) > \frac{16r^2}{\pi} e$.

Step 2. $\lim_n \int_{|x|\ge R}(e^{4\pi u_n^2} - 1)dxdy \le \frac{16r^2}{\pi} < M(\Sigma)$ and hence vanishing cannot occur.

Step 3. Dichothomy cannot occur, and hence $u \ne 0$ and $M(\Sigma)$ is achieved.

The claim in Step 1 (i.e., *concentration cannot occur*) holds true in any Ω simply connected. We sketch the proof. By contradiction, let u_n concentrate, and then $M(\Omega) = Cl(\Omega)$. So, $\exists\, \delta < 1$ and $v \in H^1_{0,\mathrm{rad}}(D)$ s. t.

$$M(\Omega) = cR^2(\Omega)Cl(D) = cR^2(\Omega)\,\delta M(D) = cR^2(\Omega)\,\delta \int_D \left(e^{4\pi v^2} - 1\right) dx.$$

Since Ω is simply connected, $\exists x_\epsilon \in \Omega$, $\exists \varphi_\epsilon$, a conformal diffeomorphism, such that $\varphi_\epsilon(D) = \Omega$, $\varphi_\epsilon(0) = x_\epsilon$ and $cR^2(\Omega) \le [\epsilon + cr_\Omega^2(x_\epsilon)] = [\epsilon + |\det J_{\varphi_\epsilon}(0)|]$. Finally, using the subharmonicity of $|\det J_{\varphi_\epsilon}(z)|$we find

$$\begin{aligned}
\delta^{-1}M(\Omega) + O(\epsilon) &= |\det J_{\varphi_\epsilon}(0)| \int_D \left(e^{4\pi v^2} - 1\right) \\
&\le 2\pi \int_0^1 r\left(e^{4\pi v^2} - 1\right)\left(\frac{1}{2\pi}\int_0^{2\pi} |\det J_{\varphi_\epsilon}(r\cos\theta, r\sin\theta)|d\theta\right) dr \\
&= \int_D \left(e^{4\pi v^2} - 1\right) |\det J_{\varphi_\epsilon}(z)dz = \int_\Omega \left(e^{4\pi(v\circ\varphi_\epsilon^{-1})^2} - 1\right) dx \le M(\Omega).
\end{aligned}$$

To rule out vanishing we need a sharp *upper bound for vanishing levels for* $M(u, \Sigma)$: this is the content of Step 2; here we need (5.11) to gain some compactness, which in turn allows to relate vanishing levels to $\lambda_1(\Sigma)$:

$$u_n = u_n^* \rightharpoonup 0 \quad \text{and in} \quad H^1_{\mathrm{loc}}(\Sigma) \quad \Rightarrow \quad \int_\Sigma \left(e^{4\pi u_n^2} - 1 - 4\pi u_n^2\right) \to 0. \tag{5.12}$$

Hence vanishing would imply

$$M(\Sigma) + \circ(1) = \int_\Sigma \left(e^{4\pi u_n^2} - 1\right) dx = \circ(1) + 4\pi \int_\Sigma u_n^2 dx \le \circ(1) + \frac{16r^2}{\pi}.$$

Actually, the two claims contained in Step 3, i.e.,

(i) *dichothomy cannot occur* (ii) $u \ne 0 \Rightarrow M(\Sigma)$ is achieved

hold true in any conformal disc $(D, \rho h)$ with ρ bounded as shown in the section below.

5.2. Extremals for Moser functionals on some conformal discs uniformly conformally equivalent to (D, h)

We consider here metrics $g = \rho h$ with $1 \le \rho \le c$ and $\rho \equiv 1$ on ∂D. Let

$$\lambda_1(g) := \inf \left\{ \int_D |\nabla u|^2 dx : \int_D u^2 dv_g = 1 \right\}.$$

Theorem 5.1. *Assume $\lambda_1(g) < \frac{1}{4}$ and*

$$\rho(0) \ge \rho(x) \ge \rho(0)(1 - |x|^2)^2 \qquad \forall x \in D.$$

Then

$$M = M(D, g) := \sup_{u \in H_0^1(D), \int_D |\nabla u|^2 \le 1} \int_D \left(e^{4\pi u^2} - 1 \right) dv_g$$

is achieved.

Proof. Let $\int_D |\nabla u_n|^2 = 1$, $u_n \rightharpoonup u$, $\int_D \left(e^{4\pi u_n^2} - 1 \right) dv_g \to M$. Then

Fact 1. $u \ne 0 \quad \Rightarrow \quad \int_D \left(e^{4\pi u^2 / \int_D |\nabla u|^2} - 1 \right) dv_g = M.$

Fact 2. $u = 0 \quad \Rightarrow \quad u_n \to 0$ in H^1_{loc} and $\int_{|x| \ge r} \left(e^{4\pi u_n^2} - 1 \right) dv_g \to M \ \forall r < 1$, i.e., only vanishing can occur.

Proof of Fact 1. If $\int_D |\nabla u|^2 = 1$, we have $u_n \to u$ in H_0^1 and hence

$$\int_D \left(e^{4\pi u_n^2} - 1 \right) dv_g \to \int_D \left(e^{4\pi u^2} - 1 \right) dv_g.$$

If

$$\alpha := \int_D |\nabla u|^2 dx < 1, \quad \beta_n := \int_D |\nabla (u_n - u)|^2 dx = 1 - \alpha + \circ(1),$$

we get, by convexity,

$$\begin{aligned}
M + \circ(1) &= \int_D \left(e^{4\pi u_n^2} - 1 \right) dv_g \\
&= \int_D \left(e^{\beta_n [\frac{\sqrt{4\pi}}{\sqrt{\beta_n}} (u_n - u)]^2 + (1-\beta_n)[\frac{4\pi}{1-\beta_n} u^2 + \frac{8\pi}{1-\beta_n} u(u_n - u)]} - 1 \right) dv_g \\
&\le \beta_n \int_D \left(e^{4\pi \frac{(u_n - u)^2}{\beta_n}} - 1 \right) dv_g + (1 - \beta_n) \int_D \left(e^{\frac{4\pi}{1-\beta_n} u^2 + \frac{8\pi}{1-\beta_n} u(u_n - u)} - 1 \right) dv_g \\
&\le [1 - \alpha + \circ(1)] M + (\alpha + \circ(1)) \left[\int_D \left(e^{4\pi (\frac{u}{\sqrt{\alpha}})^2} - 1 \right) dv_g + \circ(1) \right]
\end{aligned}$$

and hence

$$M \le (1 - \alpha) M + \alpha \int_D \left(e^{4\pi (\frac{u}{\sqrt{\alpha}})^2} - 1 \right) dv_g.$$

Proof of Fact 2. From Lions' Lemma it follows that, either

$$\int_{|x| \ge r} \left(e^{4\pi u_n^2} - 1 \right) dv_g \to M \quad \forall r < 1 \quad \text{or} \quad \exists x_0 \in \overline{D} : |\nabla u_n|^2 dx \rightharpoonup \delta_{x_0}.$$

We want to show that $|x_0| < 1$ cannot occur. Assuming $|x_0| < 1$, we find

$$\int_{d_h(x,x_0)\geq\epsilon,|x|\leq 1-\epsilon} \left(e^{4\pi u_n^2} - 1\right) dv_g \to_n 0 \quad (d_h = \text{hyperbolic distance}),$$

and so, $\forall \epsilon > 0$,

$$\lim_n \int_{|x|\geq 1-\epsilon} \left(e^{4\pi u_n^2} - 1\right) dv_g = \lambda M$$

$$\lim_n \int_{d_h(x,x_0)\leq\epsilon} \left(e^{4\pi u_n^2} - 1\right) dv_g = (1-\lambda) M$$

for some $\lambda \in [0,1]$ ($\lambda = 0$ means concentration, $0 < \lambda < 1$ means splitting). Now, choose $\varphi \in C_0^\infty(\{d_h(x,x_0) \leq \epsilon\})$ with $\varphi(x_0) = 1$ such that $\int_D |\nabla(1-\varphi)u_n|^2 = \circ(1)$. Then

$$\int_D \left(e^{4\pi(1-\varphi)^2 u_n^2} - 1\right) dv_g = \circ(1)$$

and

$$\lambda M + \circ(1) = \int_{|x|\geq 1-\epsilon} \left(e^{4\pi u_n^2} - 1\right) dv_g \leq \int_D \left(e^{4\pi(1-\varphi)^2 u_n^2} - 1\right) dv_g = \circ(1).$$

So, $\lambda = 0$ and, for a conformal diffeomorfism φ with $\varphi(0) = x_0$, and $|z_{n,\epsilon}| \leq \epsilon$,

$$\begin{aligned} M + \circ(1) &= \int_{d_h(x,x_0)\leq\epsilon} \left(e^{4\pi u_n^2} - 1\right) \frac{4\rho}{(1-|x|^2)^2} dx \\ &= \frac{4\rho(\varphi(z_{n,\epsilon}))}{(1-|z_{n,\epsilon}|^2)^2} \int_{d_h(z,0)\leq\epsilon} \left(e^{4\pi(u_n\circ\varphi)^2} - 1\right) dz. \end{aligned}$$

Now, let

$$M(D) := \int_D (e^{4\pi v^2} - 1) dz, \quad \int_D |\nabla v|^2 = 1.$$

We find

$$M = \lim_n [M + \circ(1)] \leq 4(1 + \circ_\epsilon(1))\rho(x_0) \int_D \left(e^{4\pi v^2} - 1\right) dz \quad \forall \epsilon > 0$$

and so

$$M \leq 4\rho(x_0) \int_D \left(e^{4\pi v^2} - 1\right) dz < \int_D \left(e^{4\pi v^2} - 1\right) \frac{4\rho(x)}{(1-|x|^2)^2} dx \leq M.$$

Conclusion. Let us show that if M is not achieved then $M \leq 16\pi$. To see this, let $r \approx 1$. It results

$$\begin{aligned} M &= (1+\circ(1)) \int_{|x|\geq r} \left(e^{4\pi u_n^2} - 1\right) \rho d\mu_h = (1+\circ(1)) \int_{|x|\geq r} \left(e^{4\pi u_n^2} - 1\right) d\mu_h \\ &= (1+\circ(1)) \int_D \left(e^{4\pi u_n^2} - 1\right) d\mu_h = (1+\circ(1)) \int_D \left(e^{4\pi (u_n^*)^2} - 1\right) d\mu_h. \end{aligned}$$

So, u_n^* is maximizing and hence, by Facts 1–2, it has to vanish:

$$M = \int_{|x|\geq r} \left(e^{4\pi (u_n^*)^2} - 1\right) dv_g + o(1).$$

Decay of u_n^* at $\partial\Omega$ gives

$$\int_{|x|\geq r} \left(e^{4\pi (u_n^*)^2} - 1 - 4\pi (u_n^*)^2\right) dv_g = o(1)$$

and hence

$$M = \int_D \left(e^{4\pi (u_n^*)^2} - 1 - 4\pi (u_n^*)^2\right) d\mu_h + 4\pi \int_D (u_n^*)^2 d\mu_h + o(1) \leq o(1) + 16\pi.$$

Now the conclusion follows from the following known fact (see, e.g., [5], [36]) which will imply that $M > 16\pi$.

Lemma 5.2. *Let $g = \rho g_h$ with $\rho \equiv 1$ on ∂D and $\lambda_1 = \lambda_1(g) < \frac{1}{4}$. Then there is $\varphi_1 \in H_0^1(D)$ such that $-\Delta_g \varphi_1 = \lambda_1 \varphi_1$.*

A "vanish or converge" proof. Let $\int_D |\nabla u_n|^2 = 1$, $u_n \rightharpoonup u$, $\int u_n^2 dv_g \to \frac{1}{\lambda_1}$. Assume $u = 0$. Then, for every $r < 1$ we have

$$\frac{1}{\lambda_1} + o(1) = \int_{|x|\geq r} u_n^2 dv_g = \rho(x_\epsilon) \int_{|x|\geq r} u_n^2 d\mu_h \leq 4\rho(x_\epsilon).$$

Hence $u \neq 0$.

If $u_n \to u$ in $L^2(v_g)$, then $\varphi_1 = u$. If not, $\alpha := \lim \int_D |\nabla u_n|^2 < 1$ and $\beta_n := \int_D |\nabla(u_n - u)|^2 = 1 - \alpha + o(1)$ and

$$\begin{aligned} 1/\lambda_1 + o(1) &= \int_D [\beta_n \left(\frac{u_n - u}{\sqrt{\beta_n}}\right)^2 + u^2 + 2u(u_n - u)] dv_g \\ &\leq \frac{1-\alpha}{\lambda_1} + \alpha \int_D \left(\frac{u}{\sqrt{\alpha}}\right)^2 dv_g + o(1) \end{aligned}$$

and this proves the lemma. Now, Lemma 5.2 gives

$$M \geq \int_D \left(e^{4\pi\varphi_1^2} - 1\right) dv_g \geq 4\pi \int_D \varphi_1^2 dv_g = \frac{4\pi}{\lambda_1(g)} > 16\pi. \qquad \square$$

References

[1] S. Adachi and K. Tanaka, *Inequalities in $\mathbb{R}^N$ and their best exponents*, Proc. A.M.S., Vol. 128, no. 7 (2000) 2051–2057

[2] Adams, D., *A sharp inequality of J. Moser for higher order derivatives*, Ann. of Math. 128 (1988), 385–398

[3] Adimurthi and K. Tintarev, *On a version of Trudinger–Moser Inequality with Möbius shift invariance*, Calc. Var. Partial Differential Equations 39 (2010)

[4] Adimurthi and K. Tintarev, *On compactness in the Trudinger–Moser inequality*, Annali della Scuola Normale Superiore di Pisa, accepted, DOI 10.2422/2036-2145.201111_003, arXiv:1110.3647.

[5] Agmon, S., *On the spectral theory of the Laplacian on noncompact hyperbolic manifolds*, Journées Equations aux dérivées partielles (1987) 1–16

[6] T. Aubin, *Probleme isopérimetrique et espace de Sobolev*, Journal of Differential Geometry, 11 (1976) 573–598

[7] H. Bahouri, M. Majdoub, N. Masmoudi, *On the lack of compactness in the* 2D *critical Sobolev embedding*, J. Funct. Anal. 260 (2011), no. 1, 208–252

[8] K. Bandle, M. Flucher, *Harmonic Radius and Concentration of Energy; Hyperbolic Radius and Liouville's Equations* $\Delta U = e^U$ *and* $\Delta U = U^{\frac{n+2}{n-2}}$, SIAM Review, Vol. 38, No. 2 (1996), pp. 191–238

[9] W. Beckner, *Sharp Sobolev inequalities on the sphere and the Moser–Trudinger inequality*, Ann. of math. 138 (1993), 213–242

[10] Baernstein, Albert, II *A unified approach to symmetrization*, Partial differential equations of elliptic type (Cortona, 1992), 47–91, Sympos. Math., XXXV, Cambridge Univ. Press, Cambridge, 1994.

[11] L. Battaglia, *Sobolev inequalities in the limiting case and exponential integrability*, Master Thesis, University of Roma Tre, 2011

[12] R.D. Benguria, R.L. Frank and M. Loss, *The sharp constant in the Hardy–Sobolev–Maz'ya inequality in the three dimensional upper half-space*, Math. Res. Lett. 15, no. 4, 613–622 (2008)

[13] Cao, D.M., *Nontrivial solutions of semilinear elliptic equations with critical exponent in* $\mathbb{R}^2$, Comm. Partial Diff. Equations, 17 (1992) 407–435

[14] Carleson, Chang, *On the existence of an extremal function for an inequality of J. Moser*, Bull. Sc. Math., 1986

[15] G. Carron, *Inégalitées isopérimetriques de Faber–Krahn et conséquences*, Publications de l'Institut Fourier, 220, (1992)

[16] Castorina, D.; Fabbri, I.; Mancini, G.; Sandeep, K. *Hardy–Sobolev extremals, hyperbolic symmetry and scalar curvature equations.* J. Differential Equations 246 (2009), no. 3, 1187–1206

[17] S.Y.A. Chang *Partial Differential Equations in Conformal Geometry*, Arkansas Spring Lecture Series N. 33: (2008)

[18] A. Chang, P. Yang, *Prescribing Gaussian curvature on* S^2, Acta Math. 159 (3-4), 215–259 (1987)

[19] A. Chang, P. Yang, *The inequality of Moser and Trudinger and applications to conformal geometry*, Comm. Pure and Applied Math., Vol. LVI, no. 8, August 2003, pp. 1135–1150

[20] W. Chen and C. Li, *Classification of solutions of some nonlinear elliptic equations*, Duke Math. J. 63 (1991), no. 3, 615–622

[21] H. Christianson and J.L. Marzuola, *Existence and stability of solitons for the nonlinear Schrödinger equation on Hyperbolic space*, Nonlinearity, 23(1) 89–106, 2010

[22] J. Dolbeault, *Sobolev and Hardy–Littlewood–Sobolev inequalities: duality and fast diffusion*, arXiv:1103.1145, 6 Mar. 2011

[23] Fabbri, I.; Mancini, G.; Sandeep, K.; *Classification of solutions of a critical Hardy–Sobolev operator.* J. Differential Equations 224 (2006), no. 2, 258–276

[24] Flucher, M., *Extremal functions for the Moser–Trudinger inequality in 2 dimensions.* Comm. Math. Helv., 67 (1992) 471–497

[25] L. Fontana, *Sharp borderline Sobolev inequalities on compact Riemannian manifolds,* Comment. Math. Helvetici 68 (1993) 415–454

[26] E. Hebey, *Optimal Sobolev inequalities on complete Riemannian manifolds with Ricci curvature bounded from below and positive injectivity radius,* American Journal of Mathematics, 118, (1996), 291–300

[27] E. Hebey, *Nonlinear Analysis on manifolds: Sobolev Spaces and inequalities,* Courant Lectures Notes in Math., (1999)

[28] E. Hebey, M. Vaugon, *The best constant problem in the Sobolev embedding theorem for complete Riemannian manifolds,* Duke Mathematical Journal, 79, (1995), 235–279

[29] B. Kawohl, M. Lucia, *Best constants in some exponential Sobolev inequalities,* Indiana Univ. Math. Journ., vol. 57, 4, (2008) 1907–1927

[30] M. Kwong, Kam-Li and Yi, *Uniqueness of radial solutions of semilinear elliptic equations,* Trans. Amer. Math. Soc. 333 (1992), 339–363.

[31] Li, YuXiang, *Extremal functions for Moser–Trudinger inequalities on compact Riemannian manifolds,* Sci. China Ser. A, 48, 618–648, 2005

[32] Lions, P.L., *The concentration compactness principle in the calculus of variations, The limit case, Part* 1, Rev. Mat. Iberoamericana 1, 145–201 (1985)

[33] Mancini Gabriele, *Moser–Trudinger inequality and applications to a geometric problem,* Master Thesis, University of Roma Tre, 2011

[34] Mancini, G., Sandeep, K., *On a semilinear elliptic equation in* $\mathbb{H}^n$, Ann. Sc. Norm. Super. Pisa Cl. Sci. (5) 7 (2008), no. 4, 635–671

[35] Mancini, G., Sandeep, K., *Moser–Trudinger inequalities on conformal discs,* Communications in Contemporary Mathematics Vol. 12, No. 6 (2010) 1055–1068

[36] M. Marcus, V.J. Mizel and Y. Pinchover, *On the best constant for Hardy's inequality in* $\mathbb{R}^n$, Trans. Amer. Math. Soc., Vol. 350, 8, 1998, 3237–3255

[37] Maz'ja, Vladimir G., *Sobolev Spaces,* Springer Ser. Soviet Math., Springer-Verlag, Berlin, 1985

[38] J. Moser, *On a non linear problem in differential geometry,* Dynamical Systems, Bahia 1971, Academic Press 1973

[39] J. Moser, *A sharp form of an inequality by N. Trudinger,* Indiana Univ. Math. J. 20 (1970/71), 1077–1092

[40] F. Mugelli and G. Talenti, *Sobolev Inequalities in* 2D *Hyperbolic Space. A Borderline Case,* J. of ln. equal. and Appl., 1998, Vol. 2, pp. 195–228

[41] Onofri, E., *On the positivity of the effective action in a theory of random surfaces,* Comm. Math. Phys. 6 (1982), 321–326

[42] Osserman, R., *A note on Hayman's Theorem on the bass note of a drum,* Comm. Math. Helv. 52 (1977), 545–555

[43] Osserman, R., *Isoperimetric inequalities and eigenvalues of the Laplacian,* Proc. Int. Congr, Math., Helsinki 1978

[44] X. Ren, J. Wei, *On a two dimensional elliptic problem with large exponent in nonlinearity,* Trans. Amer. Math. Soc. 343 (1994), 749–763

[45] B. Ruf, *A sharp Trudinger–Moser type inequality for unbounded domains in* $\mathbb{R}^2$, J. Funct. Anal. 219 (2005), no. 2, 340–367

[46] R. Schoen, S.T. Yau, *Conformally flat manifolds, Kleinian groups and scalar curvature*, Inventiones Mathematicae, 98, 1988, 47–71

[47] A. Tertikas and K. Tintarev, *On existence of minimizers for the Hardy–Sobolev–Maz'ya inequality*, Ann. Mat. Pura Appl. (4) 186 (2007), 645–662

[48] N.S. Trudinger, *On imbeddings into Orlicz spaces and some applications*, J. Math. Mech. **17** (1967), 473–483

Gianni Mancini
Università Roma tre
Roma, Italy
e-mail: mancini@mat.uniroma3.it

Kunnath Sandeep
TIFR-CAM
Bangalore, India
e-mail: sandeep@math.tifrbng.res.in

Concentration Analysis and Applications to PDE
Trends in Mathematics, 69–83

The Ljapunov–Schmidt Reduction for Some Critical Problems

Angela Pistoia

Abstract. We study a class of problems which are perturbation of some critical problems, namely

- the Brezis-Nirenberg problem

$$-\Delta u = |u|^{2^*-2}u + \epsilon u \text{ in } \Omega, \quad u = 0 \text{ on } \partial\Omega,$$

- the almost-critical problem

$$-\Delta u = |u|^{2^*-2-\epsilon}u \text{ in } \Omega, \quad u = 0 \text{ on } \partial\Omega,$$

- the Coron problem

$$-\Delta u = |u|^{2^*-2}u \text{ in } \Omega \setminus B(x_0, \epsilon), \quad u = 0 \text{ on } \partial\left(\Omega \setminus B(x_0, \epsilon)\right),$$

where Ω is an open bounded domain in $\mathbb{R}^n$, $n \geq 3$, $2^* = \frac{2n}{n-2}$, ϵ is a small real parameter and $x_0 \in \Omega$. In particular, we prove existence and multiplicity of positive and sign changing solutions which blow-up or blow-down at one or more points of the domain as the parameter ϵ goes to zero. The main tool is the Ljapunov–Schmidt reduction method.

Mathematics Subject Classification (2010). 35B40, 35J20, 35J55.

Keywords. Critical problems, blowing-up solutions, Ljapunov-Schmidt reduction.

1. Introduction

Let us consider the problem

$$\begin{cases} -\Delta u = |u|^{q-1}u & \text{in } \Omega, \\ u = 0 & \text{on } \partial\Omega, \end{cases} \tag{1}$$

where Ω is a smooth bounded domain in $\mathbb{R}^n$, $n \geq 3$ and $q > 1$. Let 2^* denote the critical exponent in the Sobolev embeddings, i.e., $2^* = \frac{2n}{n-2}$.

In the subcritical case, i.e., $q < 2^* - 1$, compactness of Sobolev's embedding ensures existence of at least one positive solution and infinitely many sign changing solutions to (1).

In the critical case or in the supercritical case, i.e., $q \geq 2^* - 1$, existence of solutions is a delicate issue. In [39] Pohožaev proved that the problem (1) does not admit a nontrivial solution if Ω is star-shaped. On the other hand, Kazdan and Warner in [29] proved that problem (1) has one positive radial solution and infinitely many sign changing radial solutions if Ω is an annulus. In the critical case, i.e., $q = 2^* - 1$, Bahri and Coron in [2] found a positive solution to (1) provided the domain Ω has a *nontrivial topology.*

In this survey we are in particular interested in the following perturbed critical problems.

- *The Brezis–Nirenberg problem*

$$(\mathcal{BN})_\epsilon \qquad \begin{cases} -\Delta u = |u|^{2^*-2}u + \epsilon u & \text{in } \Omega \subset \mathbb{R}^n,\ n \geq 4, \\ u = 0 & \text{on } \partial\Omega \end{cases}$$

- *The almost-critical problem*

$$(\mathcal{AC})_\epsilon \qquad \begin{cases} -\Delta u = |u|^{2^*-2-\epsilon}u & \text{in } \Omega, \\ u = 0 & \text{on } \partial\Omega \end{cases}$$

- *The Coron's problem*

$$(\mathcal{C})_\epsilon \qquad \begin{cases} -\Delta u = |u|^{2^*-2}u & \text{in } \Omega_\epsilon := \Omega \setminus B(z_0, \epsilon), \\ u = 0 & \text{on } \partial\Omega_\epsilon. \end{cases}$$

In the first two problems $\epsilon \in \mathbb{R}$ is a small parameter either positive or negative. In the last problem $z_0 \in \Omega$ and ϵ is a small positive parameter.

The common feature of those problems is that when ϵ is small enough they can have solutions u_ϵ whose shape resembles the sum of a finite number of bubbles as ϵ goes to zero, i.e.,

$$u_\epsilon(x) \sim \sum_{i=1}^{k} \lambda_i U_{\delta_i^\epsilon, z_i^\epsilon}(x)$$

where the concentration points z_i^ϵ converge to a point z_i^0 in Ω and the concentration parameters δ_i^ϵ go to zero as ϵ go to zero. If $\lambda_i = +1$ ($\lambda_i = -1$) we say that u_ϵ has a positive (negative) blow-up point at z_i^0 as ϵ goes to zero.

A *bubble* is a function

$$U_{\delta,z}(x) := \alpha_n \frac{\delta^{\frac{n-2}{2}}}{(\delta^2 + |x-z|^2)^{\frac{n-2}{2}}}, \qquad \delta > 0, \quad x, z \in \mathbb{R}^n. \tag{2}$$

Here $\alpha_n := [n(n-2)]^{\frac{n-2}{4}}$. They are positive solutions to the limit problem (see Aubin [1], Caffarelli–Gidas–Spruck [11], Talenti [44])

$$-\Delta u = u^{\frac{n+2}{n-2}} \quad \text{in } \mathbb{R}^n. \tag{3}$$

The location of the points z_i's where blowing-up occurs is strictly related to the geometry of the domain, namely Green's and Robin's functions. Let G the

Green's function of the negative Laplacian on Ω with Dirichlet boundary conditions and let H its regular part, i.e.,

$$H(x,y) = \frac{c_n}{|x-y|^{n-2}} - G(x,y), \quad \forall (x,y) \in \Omega^2,$$

where c_n is a positive constant. The function $\tau(x) := H(x,x)$, $x \in \Omega$ is called *Robin's function.* It is known that τ is a C^2-function and also that $\tau(x)$ goes to $+\infty$ as x approaches the boundary of Ω. Therefore, Robin's function has always a minimum point in Ω.

Let us state the main results concerning solutions to $(\mathcal{BN})_\epsilon$, $(\mathcal{AC})_\epsilon$ and $(\mathcal{C})_\epsilon$ which blow-up at one or more points of the domain as the parameter ϵ goes to zero.

The Brezis–Nirenberg problem and the almost critical problem when ϵ is positive

Brezis and Nirenberg in [10] proved that if $n \geq 4$ for ϵ small enough problem $(\mathcal{BN})_\epsilon$ has a positive solution. On the other hand, it is clear that the slightly sub-critical problem $(\mathcal{AC})_\epsilon$ has always a positive solution. Han in [27] proved that these solutions blow-up at a critical point of the Robin's function as ϵ goes to zero. Conversely, Rey in [40, 41] proved that any C^1-stable critical point z_0 of the Robin's function generates a family of solutions which blows-up at z_0 as ϵ goes to zero. Musso–Pistoia in [35] and Bahri–Li–Rey in [3] studied existence of solutions which blow-up at κ different points of Ω. Grossi-Takahashi [26] proved the nonexistence of positive solutions blowing up at $\kappa \geq 2$ points for these problems in convex domains.

As far as it concerns the existence of sign changing solutions, the slightly sub-critical problem $(\mathcal{AC})_\epsilon$ has infinitely many sign changing solutions. Existence of sign changing solution for problem $(\mathcal{BN})_\epsilon$ is a more difficult problem. The first result about problem $(\mathcal{BN})_\epsilon$ is due to Cerami–Solimini–Struwe, who showed in [13] the existence of a pair of least energy sign changing solutions if $n \geq 6$ and ϵ is small enough. The existence of infinitely many solutions to $(\mathcal{BN})_\epsilon$ for any $\epsilon > 0$ was established by Devillanova–Solimini in [22] when $n \geq 7$. Moreover, for low dimensions $n = 4, 5, 6$, in [23] they proved the existence of at least $n+1$ pairs of solutions provided ϵ is small enough. Ben Ayed–El Mehdi–Pacella in [7, 8] studied the blow up of the low energy sign-changing solutions of problems $(\mathcal{AC})_\epsilon$ and $(\mathcal{BN})_\epsilon$ as ϵ goes to zero and they classified these solutions according to the concentration speeds of the positive and negative part. In [12] Castro–Clapp proved the existence of one pair of solutions in a symmetric domain which change sign exactly once, provided $n \geq 4$ and ϵ is small enough. Moreover they describe the profile of the solutions, by showing that the solutions blow-up positively and negatively at two different points in Ω as ϵ goes to 0. Micheletti–Pistoia in [31] and Bartsch–Micheletti–Pistoia in [4] generalized such a result showing the existence of at least n pairs of sign changing solutions with one negative and one positive blow-up points. Pistoia–Weth in [38] and Musso–Pistoia in [36] proved that a large number of sign changing solutions exists for problem $(\mathcal{AC})_\epsilon$: the solutions are a superposition with alternating sign of bubbles whose centers collapse to the minimum point of the Robin's function as ϵ goes to zero. This result is unknown for problem $(\mathcal{BN})_\epsilon$ even if we think to be true.

The Brezis–Nirenberg problem and the almost critical problem when ϵ is negative

In [39] Pohožaev proved that the problems $(\mathcal{B})_\epsilon$ and $(\mathcal{AC})_\epsilon$ do not have any solutions if ϵ is negative and Ω is starshaped.

As far as it concerns the existence of blowing-up solutions, when ϵ is negative and small enough, a completely different phenomena take place even if the domain Ω is not starshaped. Indeed, Ben Ayed–El Mehdi–Grossi–Rey in [6] proved that problem $(\mathcal{AC})_\epsilon$ do not have any positive solutions which blows-up at one point when ϵ goes to zero. We believe that their argument could also be extended to the problem $(\mathcal{BN})_\epsilon$. Del Pino–Felmer–Musso in [21] and Musso–Pistoia in [32] found, for ϵ small enough, a positive solutions with two positive blow-up points provided the domain Ω has a hole. Del Pino–Felmer–Musso in [19, 20] and Pistoia–Rey in [37] found solutions with three or more positive blow-up points, under suitable assumptions on the domain Ω. Towers of positive bubbles were constructed by del Pino–Dolbeault–Musso in [17, 18] and by Ge–Jing–Pacard in [25], under suitable assumptions on non degeneracy of Robin's and Green's functions. As far as it concerns the study of sign changing solutions, Ben Ayed–Bouh in [5] proved that problem $(\mathcal{AC})_\epsilon$, does not have any sign changing solutions with one positive and one or two negative blow-up points. We believe that their argument could also be extended also to the problem $(\mathcal{BN})_\epsilon$. There are no results about existence of sign changing solutions for these problems.

The Coron problem

Coron in [16] found via variational methods a solution to problem $(\mathcal{C})_\epsilon$ provided ϵ is small enough. If the domain has several holes, Rey in [42] and Li–Yan–Yang in [30] constructed solutions blowing-up at the centers of the holes as the size of the holes goes to zero. On the other hand, Clapp–Weth in [15] found a second solution to $(\mathcal{C})_\epsilon$, but they were unable to say if it was positive or changed sign. Clapp–Musso–Pistoia in [14] found positive and sign changing solutions to $(\mathcal{C})_\epsilon$ blowing-up at the center of the hole and at one or more points inside the domain as ϵ goes to zero. If the domain has two small holes, Musso–Pistoia in [34] constructed a sign changing solution with one positive blow-up point and one negative blow-up point at the centers of the two holes. Musso–Pistoia in [33] and Ge–Musso–Pistoia in [24] found a large number of sign changing solutions to $(\mathcal{C})_\epsilon$: the solutions are a superposition of bubbles with alternating sign whose centers collapse to the center of the hole as ϵ goes to zero.

The proofs of all the results concerning existence of solutions which blow-up positively or negatively at one or more points as the parameter ϵ goes to zero, rely on a Lyapunov–Schmidt reduction scheme firstly developed by Bahri–Coron in [2]. This allows to reduce the problem of finding blowing-up solutions to the problem of finding critical points of a functional which depends only on the blow-up points and the concentration rates. The leading part of the reduced functional is explicitly given in terms of the geometry of the domain, namely Green's and Robin's functions. The reduced functional also takes into account the different interactions among the bubbles which depends on their respective sign. Finally, we use a variational approach and we obtain the existence of critical points of the reduced functional by applying a minimization argument or a min-max argument. In the following we describe the main steps to get some of the previous results. We will refer to [35] and [3, 21, 31] for the proofs related to the construction of positive and sign-changing multi-bubbles to problems $(\mathcal{BN})_\epsilon$ and $(\mathcal{AC})_\epsilon$, respectively. We will refer to [36] and to [33, 24] for the proofs related to the construction of towers of bubbles to problems $(\mathcal{AC})_\epsilon$ and $(\mathcal{C})_\epsilon$, respectively.

2. Setting of the problem

We want to rewrite problems $(\mathcal{BN})_\epsilon$, $(\mathcal{AC})_\epsilon$ and $(\mathcal{C})_\epsilon$ in a different, but equivalent, form.

Let us take

$$(u,v) := \int_\Omega \nabla u \cdot \nabla v \, dx, \qquad \|u\| := \left(\int_\Omega |\nabla u|^2 \, dx \right)^{1/2},$$

as the inner product in $\mathrm{H}_0^1(\Omega)$ and its corresponding norm. Similarly, for each $r \in [1, \infty)$,

$$\|u\|_r := \left(\int_\Omega |u|^r \, dx \right)^{1/r}$$

is a norm in $\mathrm{L}^r(\Omega)$.

Let $i^* : \mathrm{L}^{\frac{2n}{n+2}}(\Omega) \to \mathrm{H}_0^1(\Omega)$ be the adjoint operator to the embedding $i : \mathrm{H}_0^1(\Omega) \hookrightarrow \mathrm{L}^{\frac{2n}{n-2}}(\Omega)$, i.e., $i^*(u) = v$ if and only if

$$(v, \varphi) = \int_\Omega u(x)\varphi(x) dx \quad \text{for all } \varphi \in C_c^\infty(\Omega)$$

if and only if

$$-\Delta v = u \quad \text{in } \Omega, \qquad v = 0 \quad \text{on } \partial\Omega.$$

It is clear that i^* is a continuous map, namely there exists a positive constant c such that

$$\|i^*(u)\| \le c \, \|u\|_{\frac{2n}{n+2}} \qquad \forall \, u \in \mathrm{L}^{\frac{2n}{n+2}}(\Omega).$$

To study the slightly supercritical case, namely problem $(\mathcal{AC})_\epsilon$ with $\epsilon < 0$, we need to find solutions in the space $\mathrm{H}_0^1(\Omega) \cap \mathrm{L}^{s_\epsilon}(\Omega)$ with $s_\epsilon := \frac{2n}{n-2} - \epsilon\frac{n}{2}$. Indeed, by a well-known Hardy–Littlewood–Sobolev inequality (see Hardy–Littlewood [28] and Sobolev [45]) we deduce that i^* restricted to $\mathrm{H}_0^1(\Omega) \cap \mathrm{L}^{s_\epsilon}(\Omega)$ is a continuous map, namely

$$\|i^*(u)\|_s \le c \, \|u\|_{\frac{s_\epsilon}{\frac{n+2}{n-2} - \epsilon}}$$

for some positive constant c which depends only on n. We point out that if $\epsilon > 0$ then $\mathrm{H}_0^1(\Omega) \cap \mathrm{L}^{s_\epsilon}(\Omega)$ coincides with $\mathrm{H}_0^1(\Omega)$.

Using the above definitions and notations, it is clear that our problems can be rewritten in the equivalent form

$$\begin{cases} u = i^* \left[f_\epsilon(u)\right] \\ u \in \mathfrak{H}_\epsilon, \end{cases} \tag{4}$$

where

(i) *for problem* $(\mathcal{BN})_\epsilon$

$$f_\epsilon(u) := |u|^{p-1}s + \epsilon u \quad \text{and} \quad \mathfrak{H}_\epsilon := \mathrm{H}_0^1(\Omega)$$

(ii) *for problem* $(\mathcal{AC})_\epsilon$

$$f_\epsilon(u) := |u|^{p-1-\epsilon}u \quad \text{and} \quad \mathfrak{H}_\epsilon := \mathrm{H}_0^1(\Omega) \cap \mathrm{L}^{s_\epsilon}(\Omega)$$

(iii) *for problem* $(\mathcal{C})_\epsilon$

$$f_\epsilon(u) := |u|^{p-1}u \quad \text{and} \quad \mathfrak{H}_\epsilon := \mathrm{H}_0^1(\Omega_\epsilon).$$

3. The Ljapunov-Schmidt procedure

3.1. The approximating solution

The first step is writing a *good approximating solution.*

Let PW denote the projection of the function $W \in D^{1,2}(\mathbb{R}^n)$ onto $\mathrm{H}_0^1(D)$, i.e.,

$$\Delta PW = \Delta W \quad \text{in } D, \qquad PW = 0 \quad \text{on } \partial D,$$

where D is a smooth bounded domain in $\mathbb{R}^n$.

Let $\kappa \geq 1$ be a fixed integer. We look for solutions u_ϵ to problems $(\mathcal{BN})_\epsilon$, $(\mathcal{AC})_\epsilon$ and $(\mathcal{C})_\epsilon$ as

$$u_\epsilon(x) = V_{\mathbf{z},\mathbf{d}}(x) + \phi(x), \quad V_{\mathbf{z},\mathbf{d}}(x) := \sum_{i=1}^{\kappa} \lambda_i PU_{\delta_i, z_i}(x), \tag{5}$$

where the higher-order term ϕ belongs to a suitable space described in the next subsection. Here $\lambda_i \in \{-1, +1\}$, the concentration points z_i's lye in Ω and the concentration parameters δ_i's are chosen as follows.

- *Multi-bubbles for problem* $(\mathcal{BN})_\epsilon$

$$\mathfrak{M} - (\mathcal{BN})_\epsilon \qquad \begin{cases} \lambda_i \in \{-1, +1\} \\ z_1, \ldots, z_\kappa \in \Omega \text{ and } z_i \neq z_j \\ \delta_i = |\epsilon|^{\frac{1}{n-4}} d_i \quad \text{with} \quad d_i > 0 \end{cases}$$

The configuration space is

$$\Lambda := \big\{(\mathbf{z},\mathbf{d}) \ : \ \mathbf{z} = (z_1,\dots,z_\kappa) \in \Omega^\kappa, \ z_i \neq z_j, \\ \mathbf{d} = (d_1,\dots,d_\kappa) \in (0,+\infty)^\kappa\big\}.$$

- *Multi-bubbles for problem* $(\mathcal{AC})_\epsilon$

$$\mathfrak{M} - (\mathcal{AC})_\epsilon \qquad \begin{cases} \lambda_i \in \{-1,+1\} \\ z_1,\dots,z_\kappa \in \Omega \quad \text{and} \quad z_i \neq z_j \\ \delta_i = |\epsilon|^{\frac{1}{n-2}} d_i \quad \text{with} \quad d_i > 0. \end{cases}$$

The configuration space is

$$\Lambda := \big\{(\mathbf{z},\mathbf{d}) \ : \ \mathbf{z} = (z_1,\dots,z_\kappa) \in \Omega^\kappa, \ z_i \neq z_j, \\ \mathbf{d} = (d_1,\dots,d_\kappa) \in (0,+\infty)^\kappa\big\}.$$

- *Tower of bubbles with alternating sign for problem* $(\mathcal{AC})_\epsilon$ *when* $\epsilon > 0$

$$\mathfrak{T} - (\mathcal{AC})_\epsilon \qquad \begin{cases} \lambda_i = (-1)^i \\ z_i = z + \delta_i \sigma_i \in \Omega \quad \text{with} \quad \sigma_1,\dots,\sigma_{\kappa-1} \in \mathbb{R}^n \text{ and } \sigma_\kappa = 0 \\ \delta_i = \epsilon^{\frac{2(i-1)+1}{N-2}} d_i \quad \text{with } d_i > 0. \end{cases}$$

The configuration space is

$$\Lambda := \big\{(\mathbf{z},\mathbf{d}) \ : \ \mathbf{z} = (\sigma_1,\dots,\sigma_{\kappa-1},z) \in \mathbb{R}^{(\kappa-1)n} \times \Omega, \\ \mathbf{d} = (d_1,\dots,d_\kappa) \in (0,+\infty)^\kappa\big\}.$$

- *Tower of bubbles with alternating sign for problem* $(\mathcal{C})_\epsilon$

$$\mathfrak{T} - (\mathcal{C})_\epsilon \qquad \begin{cases} \lambda_i = (-1)^i \\ z_i = z_0 + \delta_i \sigma_i \in \Omega \quad \text{with} \quad \sigma_1,\dots,\sigma_k \in \mathbb{R}^N \\ \delta_i = \epsilon^{\frac{2i-1}{2k}} d_i \quad \text{with } d_i > 0. \end{cases}$$

The configuration space is

$$\Lambda := \big\{(\mathbf{z},\mathbf{d}) \ : \ \mathbf{z} = (\sigma_1,\dots,\sigma_\kappa) \in \mathbb{R}^{\kappa n}, \\ \mathbf{d} = (d_1,\dots,d_\kappa) \in (0,+\infty)^\kappa\big\}.$$

To say that $V_{\mathbf{z},\mathbf{d}}$ is a *good approximating solution* we need to estimate the error

$$\mathfrak{R}_{\mathbf{z},\mathbf{d}} := V_{\mathbf{z},\mathbf{d}} - i^* \left[f_\epsilon(V_{\mathbf{z},\mathbf{d}})\right] \in \mathfrak{H}_\epsilon.$$

Proposition 3.1. *For any compact subset* $\mathbf{C}$ *of* Λ *there exist* $\epsilon_0 > 0$ *and* $c > 0$ *such that for each* $\epsilon \in (0,\epsilon_0)$ *and* $(\mathbf{z},\mathbf{d}) \in \mathbf{C}$ *we have*

$$\|\mathfrak{R}_{\mathbf{z},\mathbf{d}}\| \leq c|\epsilon|^\eta.$$

for some $\eta > 0$ *which depends only on* n *and* κ.

3.2. The equation becomes a system

The second step is writing the equation as a system.

We need to fix the space where the rest term ϕ in (5) belongs to. It is important to introduce the functions

$$\psi^0_{\delta,z}(x) := \frac{\partial U_{\delta,z}}{\partial \delta} = \alpha_n \frac{n-2}{2} \delta^{\frac{n-4}{2}} \frac{|x-z|^2-\delta^2}{(\delta^2+|x-z|^2)^{n/2}}$$

and, for each $j = 1, \dots, n$,

$$\psi^j_{\delta,z}(x) := \frac{\partial U_{\delta,z}}{\partial z_j} = \alpha_n (n-2) \delta^{\frac{n-2}{2}} \frac{x_j - z_j}{(\delta^2+|x-z|^2)^{n/2}},$$

which span the set of solutions to the linearized problem (see Bianchi–Egnell [9])

$$-\Delta \psi = f_0'(U_{\delta,z}) \psi \text{ in } \mathbb{R}^n.$$

Remember that the δ_i's, the z_i's and the configuration space Λ are given in $\mathfrak{M}-(\mathcal{BN})_\epsilon$, $\mathfrak{M}-(\mathcal{AC})_\epsilon$, $\mathfrak{T}-(\mathcal{AC})_\epsilon$ and $\mathfrak{T}-(\mathcal{C})_\epsilon$.

If $(\mathbf{z},\mathbf{d}) \in \Lambda$, we introduce the spaces

$$K_{\mathbf{z},\mathbf{d}} := \operatorname{span}\{P\psi^j_{\delta_i,z_i} \ : \ i = 1, \dots, \kappa, \ j = 0, 1, \dots, n\},$$
$$K^\perp_{\mathbf{z},\mathbf{d}} := \left\{\phi \in \mathfrak{H}_\epsilon \ : \ (\phi, P\psi^j_{\delta_i,z_i}) = 0, \ i = 1, \dots, \kappa, \ j = 0, 1, \dots, n\right\}$$

and the projection operators

$$\Pi_{\mathbf{z},\mathbf{d}}(u) := \sum_{i=1}^{\kappa}\sum_{j=0}^{n}(u, P\psi^j_{\delta_i,z_i})P\psi^j_{\delta_i,z_i} \quad \text{and} \quad \Pi^\perp_{\mathbf{z},\mathbf{d}}(u) := u - \Pi_{\mathbf{z},\mathbf{d},}(u).$$

Our approach to solve problem (4) will be to solve the system

$$\begin{cases} \Pi^\perp_{\mathbf{z},\mathbf{d}}\Big\{V_{\mathbf{z},\mathbf{d}} + \phi - i^*\left[f_\epsilon(V_{\mathbf{z},\mathbf{d}}+\phi)\right]\Big\} = 0, \\ \Pi_{\mathbf{z},\mathbf{d}}\Big\{V_{\mathbf{z},\mathbf{d},} + \phi - i^*\left[f_\epsilon(V_{\mathbf{z},\mathbf{d},}+\phi)\right]\Big\} = 0, \\ (\mathbf{z},\mathbf{d}) \in \Lambda \quad \text{and} \quad \phi \in K^\perp_{\mathbf{z},\mathbf{d}}. \end{cases} \tag{6}$$

3.3. The reduction argument

The third step is reducing the problem to a finite-dimensional one.

The first result we need concerns the invertibility of the linear operator $\mathfrak{L}_{\mathbf{z},\mathbf{d}} : K^\perp_{\mathbf{z},\mathbf{d}} \to K^\perp_{\mathbf{z},\mathbf{d}}$ defined by

$$\mathfrak{L}_{\mathbf{z},\mathbf{d}}\phi := \phi - \Pi^\perp_{\mathbf{z},\mathbf{d},} i^*\left[f_0'(V_{\mathbf{z},\mathbf{d}})\phi\right].$$

We will prove the following.

Proposition 3.2. *For any compact subset* $\mathbf{C}$ *of* Λ *there exist* $\epsilon_0 > 0$ *and* $c > 0$ *such that for each* $\epsilon \in (0, \epsilon_0)$ *and* $(\mathbf{z},\mathbf{d}) \in \mathbf{C}$ *the operator* $L_{\mathbf{z},\mathbf{d}}$ *is invertible and*

$$\|L_{\mathbf{z},\mathbf{d}}\phi\| \geq c\,\|\phi\| \qquad \forall\, \phi \in K^\perp_{\mathbf{z},\mathbf{d}}.$$

Secondly, we solve the first equation in system (6), namely for each $(\mathbf{z},\mathbf{d}) \in \Lambda$ and small ϵ we find a function $\phi \in K^{\perp}_{\mathbf{z},\mathbf{d}},$ which solves the first equation in system (6). To do this, we use a simple contraction mapping argument together with the estimate of the rest term given in Proposition 3.1.

Proposition 3.3. *For any compact subset* $\mathbf{C}$ *of* Λ *there exist* $\epsilon_0 > 0$ *and* $c > 0$ *such that for each* $\epsilon \in (0,\epsilon_0)$ *and* $(\mathbf{z},\mathbf{d}) \in \mathbf{C}$ *there exists a unique* $\phi^{\epsilon}_{\mathbf{z},\mathbf{d}} \in K^{\perp}_{\mathbf{s},\mathbf{d}}$ *which solves the first equation in system* (6) *and satisfies*

$$\left\|\phi^{\epsilon}_{\mathbf{z},\mathbf{d}}\right\| \leq c|\epsilon|^{\eta},$$

where η *is given in Proposition* 3.1.

Finally, we reduce the problem to a finite-dimensional one.

We introduce the energy functional $J_\epsilon : \mathfrak{H}_\epsilon \to \mathbb{R}$ defined by

$$J_\epsilon(u) := \frac{1}{2}\int\limits_{\Omega} |\nabla u|^2 dx - \int\limits_{\Omega} F_\epsilon(u),$$

where

(i) *for problem* $(\mathcal{BN})_\epsilon$

$$F_\epsilon(u) := \frac{1}{p+1}|u|^{p+1} + \frac{1}{2}\epsilon u^2 \quad \text{and} \quad \mathfrak{H}_\epsilon := H^1_0(\Omega)$$

(ii) *for problem* $(\mathcal{AC})_\epsilon$

$$F_\epsilon(u) := \frac{1}{p+1-\epsilon}|u|^{p+1-\epsilon} \quad \text{and} \quad \mathfrak{H}_\epsilon := H^1_0(\Omega) \cap L^{s_\epsilon}(\Omega)$$

(iii) *for problem* $(\mathcal{C})_\epsilon$

$$F_\epsilon(u) := \frac{1}{p+1}|u|^{p+1} \quad \text{and} \quad \mathfrak{H}_\epsilon := H^1_0(\Omega_\epsilon).$$

It is well known that critical points of J_ϵ are the solutions to problem (4). We introduce the *reduced energy* $\widetilde{J}_\epsilon : \Lambda \to \mathbb{R}$ by

$$\widetilde{J}_\epsilon(\mathbf{z},\mathbf{d}) := J_\epsilon(V_{\mathbf{z},\mathbf{d}} + \phi^{\epsilon}_{\mathbf{z},\mathbf{d}})$$

We prove that critical points of $\widetilde{J}_\epsilon$ generates solutions to the second equation in system (6) and so blowing-up solutions to problem (4).

Proposition 3.4. *The function* $V_{\mathbf{z},\mathbf{d}} + \phi^{\epsilon}_{\mathbf{z},\mathbf{d}}$ *is a critical point of the functional* J_ϵ *if and only if the point* $(\mathbf{z},\mathbf{d})$ *is a critical point of the function* $\widetilde{J}_\epsilon$.

3.4. The reduced problem

The last step is looking for critical points of the *reduced energy* $\widetilde{J}_\epsilon$.

To to this, we need an accurate asymptotic expansion of the *reduced energy* $\widetilde{J}_\epsilon$,

Proposition 3.5. *It holds true that*

$$\widetilde{J}_\epsilon(\mathbf{z},\mathbf{d}) = a(\epsilon) + b|\epsilon|^\gamma \Phi(\mathbf{z},\mathbf{d}) + o\left(|\epsilon|^\gamma\right) \tag{7}$$

C^1-uniformly on compact sets of Λ. Here $a(\epsilon)$ is constant which depends only on n, κ and ϵ and b is a constant. The positive constant γ and the function Φ are defined as follows.

In the case $\mathfrak{M}-(\mathcal{BN})_\epsilon$: $\quad \gamma := \frac{n-2}{n-4}$ *and*

$$\Phi(\mathbf{z},\mathbf{d}) := \begin{cases} \sum\limits_{i=1}^{\kappa} d_i^{n-2} H(z_i,z_i) + \sum\limits_{i=1}^{\kappa} \lambda_i\lambda_j \left(d_i d_j\right)^{\frac{n-2}{2}} G(z_i,z_j) - \sum\limits_{i=1}^{\kappa} d_i^2 & \text{if } \epsilon > 0 \\ \sum\limits_{i=1}^{\kappa} d_i^{n-2} H(z_i,z_i) + \sum\limits_{i=1}^{\kappa} \lambda_i\lambda_j \left(d_i d_j\right)^{\frac{n-2}{2}} G(z_i,z_j) + \sum\limits_{i=1}^{\kappa} d_i^2 & \text{if } \epsilon < 0 \end{cases} \tag{8}$$

In the case $\mathfrak{M}-(\mathcal{AC})_\epsilon$: $\quad \gamma := 1$ *and*

$$\Phi(\mathbf{z},\mathbf{d}) := \begin{cases} \sum\limits_{i=1}^{\kappa} d_i^{n-2} H(z_i,z_i) + \sum\limits_{i=1}^{\kappa} \lambda_i\lambda_j \left(d_i d_j\right)^{\frac{n-2}{2}} G(z_i,z_j) - \sum\limits_{i=1}^{\kappa} \ln d_i & \text{if } \epsilon > 0 \\ \sum\limits_{i=1}^{\kappa} d_i^{n-2} H(z_i,z_i) + \sum\limits_{i=1}^{\kappa} \lambda_i\lambda_j \left(d_i d_j\right)^{\frac{n-2}{2}} G(z_i,z_j) + \sum\limits_{i=1}^{\kappa} \ln d_i & \text{if } \epsilon < 0 \end{cases} \tag{9}$$

In the case $\mathfrak{T}-(\mathcal{AC})_\epsilon$: $\quad \gamma := 1$ *and*

$$\Psi(\mathbf{z},\mathbf{d}) := H(z,z)d_1^{n-2} - \sum_{i=1}^{\kappa-1} \frac{1}{(1+\sigma_i^2)^{\frac{n-2}{2}}} \left(\frac{d_{i+1}}{d_i}\right)^{\frac{n-2}{2}} - \sum_{j=1}^{\kappa} \ln d_i. \tag{10}$$

In the case $\mathfrak{T}-(\mathcal{C})_\epsilon$: $\quad \gamma := \frac{n-2}{2\kappa}$ *and*

$$\Psi(\mathbf{z},\mathbf{d}) := H(z_0,z_0)d_1^{n-2} + \frac{1}{(1+\sigma_k^2)^{n-2}} \frac{1}{d_k^{n-2}} + \sum_{j=1}^{\kappa-1} \frac{1}{(1+\sigma_j^2)^{\frac{n-2}{2}}} \left(\frac{d_{j+1}}{d_j}\right)^{\frac{n-2}{2}}. \tag{11}$$

Finally, we reduce the problem of finding blowing-up solutions to the problem (4) to the problem of finding *good* critical points of the function Ψ, which is defined on a finite-dimensional space.

Theorem 3.6. *Assume* $(\mathbf{z}^*, \mathbf{d}^*) \in \Lambda$ *is a* C^1*-stable critical points of the function* Ψ. *Then if* ϵ *is small enough there exists a solution* u_ϵ *to problem* (4) *such that*

$$\|u_\epsilon - V_{\mathbf{z}_\epsilon, \mathbf{d}_\epsilon}\| \to 0 \quad as \ \ \epsilon \to 0$$

with

$$(\mathbf{z}_\epsilon, \mathbf{d}_\epsilon) \to (\mathbf{z}^*, \mathbf{d}^*) \quad as \ \ \epsilon \to 0$$

Proof. It follows immediately from the expansion (7) and Proposition 3.4. □

We remark that $(\mathbf{z}^*, \mathbf{d}^*) \in \Lambda$ is a C^1-stable critical points of the function Ψ if

(i) $(\mathbf{z}^*, \mathbf{d}^*) \in \Lambda$ is an isolated minimum point of Ψ,
(ii) $(\mathbf{z}^*, \mathbf{d}^*) \in \Lambda$ is an isolated maximum point of Ψ,
(iii) $(\mathbf{z}^*, \mathbf{d}^*) \in \Lambda$ is a non degenerate critical point of Ψ.
(iv) $(\mathbf{z}^*, \mathbf{d}^*) \in \Lambda$ is a critical point of min-max type of Ψ (according to the Definition given by Del Pino–Felmer–Musso in [21])

4. Examples

4.1. Multi-bubbles for the Brezis–Nirenberg problem $(\mathcal{BN})_\epsilon$ and the slightly sub-critical problem $(\mathcal{AC})_\epsilon$ when ϵ is positive

According to Theorem 3.6, solutions to these problems are generated by C^1-stable critical points of the function Ψ defined in (8) and (9) when $\epsilon > 0$.

- $\kappa = 1$
  The function Ψ has a minimum point $\Rightarrow$ if $\epsilon \sim 0^+$ problems $(\mathcal{BN})_\epsilon$ and $(\mathcal{AC})_\epsilon$ have a positive solution with one blow-up point.
- $\kappa = 2$, $\lambda_1 = \lambda_2 = +1$, Ω *is a dumb-bell with a thin handle*
  The function Ψ has a minimum point $\Rightarrow$ if $\epsilon \sim 0^+$ problems $(\mathcal{BN})_\epsilon$ and $(\mathcal{AC})_\epsilon$ have a positive solution with two different blow-up points.
- $\kappa \geq 2$, $\lambda_1 = \cdots = \lambda_\kappa = +1$, Ω *is convex*
  The function Ψ does not have any critical points $\Rightarrow$ if $\epsilon \sim 0^+$ problems $(\mathcal{BN})_\epsilon$ and $(\mathcal{AC})_\epsilon$ do not have a positive solution with κ different blow-up points.
- $\kappa = 2$, $\lambda_1 = +1$, $\lambda_2 = -1$
  The function Ψ has a minimum point and $(n-1)$ critical points of min-max type $\Rightarrow$ if $\epsilon \sim 0^+$ problems $(\mathcal{BN})_\epsilon$ and $(\mathcal{AC})_\epsilon$ have n sign changing solution with one positive and one negative blow-up points.

4.2. Multi-bubbles for the Brezis–Nirenberg problem $(\mathcal{BN})_\epsilon$ and the slightly super-critical problem $(\mathcal{AC})_\epsilon$ when ϵ is negative

According to Theorem 3.6, solutions to these problems are generated by C^1-stable critical points of the function Ψ defined in (8) and (9) when $\epsilon < 0$.

- $\kappa = 1$
  The function Ψ does not have any critical points $\Rightarrow$ if $\epsilon \sim 0^-$ problems $(\mathcal{BN})_\epsilon$ and $(\mathcal{AC})_\epsilon$ do not have any positive solution with one blow point.

- $\kappa = 2$, $\lambda_1 = \lambda_2 = +1$ or $\kappa = 3$, $\lambda_1 = \lambda_2 = \lambda_3 = +1$, Ω *has a hole*
  The function Ψ has a critical point of min-max type $\Rightarrow$ if $\epsilon \sim 0^-$ problems $(\mathcal{BN})_\epsilon$ and $(\mathcal{AC})_\epsilon$ have a positive solution with two different positive blow-up points.
- $\kappa = 2$, $\lambda_1 = +1$, $\lambda_2 = -1$
  The function Ψ does not have any critical points $\Rightarrow$ if $\epsilon \sim 0^-$ problems $(\mathcal{BN})_\epsilon$ and $(\mathcal{AC})_\epsilon$ do not have any sign changing solution with one positive and one negative blow-up points.

4.3. Tower of bubbles with alternating sign for the almost critical problem $(\mathcal{AC})_\epsilon$ when ϵ is negative

According to Theorem 3.6, solutions to this problem are generated by C^1-stable critical points of the function Ψ defined in (10).

- $\kappa \geq 1$
  The function Ψ has a critical point of min-max type $\Rightarrow$ if $\epsilon \sim 0^+$ problems $(\mathcal{AC})_\epsilon$ has a sign changing solution with κ collapsing blow-up points with alternating sign.

4.4. Tower of bubbles with alternating sign for the Coron's problem $(\mathcal{C})_\epsilon$

According to Theorem 3.6, solutions to this problem are generated by C^1-stable critical points of the function Ψ defined in (11).

- $\kappa \geq 1$
  The function Ψ has a critical point of min-max type $\Rightarrow$ if $\epsilon \sim 0^+$ problems $(\mathcal{C})_\epsilon$ has a sign changing solution with κ collapsing blow-up points with alternating sign.

References

[1] T. Aubin: *Problèmes isopérimétriques et espaces de Sobolev,* J. Differential Geometry **11** (1976), 573–598.

[2] A. Bahri, J.M. Coron: *On a nonlinear elliptic equation involving the critical Sobolev exponent: the effect of the topology of the domain,* Comm. Pure Appl. Math. **41** (1988), 253–294.

[3] A. Bahri, Y. Li, O. Rey: *On a variational problem with lack of compactness: the topological effect of the critical points at infinity,* Calc. Var. Partial Differential Equations **3** (1995), 67–93.

[4] T. Bartsch, A. Micheletti, A. Pistoia: *On the existence and the profile of nodal solutions of elliptic equations involving critical growth,* Calc. Var. Partial Differential Equations **26** (2006), 265–282.

[5] M. Ben Ayed, K.O. Bouh: *Nonexistence results of sign-changing solutions to a supercritical nonlinear problem.* Commun. Pure Appl. Anal. **7** (2008), no. 5, 1057–1075

[6] M. Ben Ayed, K. El Mehdi, M. Grossi, O. Rey: *A nonexistence result of single peaked solutions to a supercritical nonlinear problem.* Commun. Contemp. Math. **5** (2003), no. 2, 179–195

[7] M. Ben Ayed, K. El Mehdi, F. Pacella: *Classification of low energy sign-changing solutions of an almost critical problem.* J. Funct. Anal. **50** (2007), no. 2, 347–373.

[8] M. Ben Ayed, K. El Mehdi, F. Pacella: *Blow-up and symmetry of sign-changing solutions to some critical elliptic equations.* J. Differential Equations **230** (2006), no. 2, 771–795

[9] G. Bianchi, H. Egnell: *A note on the Sobolev inequality,* J. Funct. Anal. **100**-(1) (1991), 18–24.

[10] H. Brézis, L. Nirenberg: *Positive solutions of nonlinear elliptic equations involving critical Sobolev exponents,* Comm. Pure Appl. Math. **36**-4 (1983), 437–477.

[11] L. Caffarelli, B. Gidas, J. Spruck: *Asymptotic symmetry and local behavior of semilinear elliptic equations with critical Sobolev growth,* Comm. Pure Appl. Math. **42** (1989), 271–297.

[12] A. Castro, M. Clapp: *The effect of the domain topology on the number of minimal nodal solutions of an elliptic equation at critical growth in a symmetric domain,* Nonlinearity **16** (2003), 579–590.

[13] G. Cerami, S. Solimini, M. Struwe: *Some existence results for superlinear elliptic boundary value problems involving critical exponents* J. Funct. Anal. **69** (1986), 298–306.

[14] M. Clapp, M. Musso, A. Pistoia: *Multipeak solutions to the Bahri–Coron problem in domains with a shrinking hole.* J. Funct. Anal. **256** (2009), no. 2, 275–306

[15] M. Clapp, T. Weth: *Minimal nodal solutions of the pure critical exponent problem on a symmetric domain.* Calc. Var. Partial Differential Equations **21** (2004), no. 1, 1–14.

[16] J.M. Coron: *Topologie et cas limite des injections de Sobolev.* C. R. Acad. Sci. Paris Ser. I Math. **299** (1984), no. 7, 209–212.

[17] M. del Pino, J. Dolbeault, M. Musso: *Bubble-tower radial solutions in the slightly supercritical Brezis–Nirenberg problem.* J. Differential Equations **193** (2003), no. 2, 280–306.

[18] M. del Pino, J. Dolbeault, M. Musso: *The Brezis–Nirenberg problem near criticality in dimension* 3. J. Math. Pures Appl. **83** (2004), no. 12, 1405–1456.

[19] M. Del Pino, P. Felmer, M. Musso: *Multi-bubble solutions for slightly super-critical elliptic problems in domains with symmetries,* Bull. London Math. Soc. **35** (2003), 513–521.

[20] M. Del Pino, P. Felmer, M. Musso: *Multi-peak solutions for super-critical elliptic problems in domains with small holes,* J. Differential Equations **182** (2002), 511–540.

[21] M. Del Pino, P. Felmer, M. Musso: *Two-bubble solutions in the super-critical Bahri–Coron's problem,* Calc. Var. Partial Differential Equations **16** (2003), 113–145.

[22] G. Devillanova, S. Solimini: *Concentration estimates and multiple solutions to elliptic problems at critical growth,* Advan. Diff. Equat. **7** (2002), 1257–1280.

[23] G. Devillanova, S. Solimini: *A multiplicity result for elliptic equations at critical growth in low dimension,* Comm. Contemp. Math. (to appear).

[24] Y. Ge, M. Musso, A. Pistoia: *Sign changing tower of bubbles for an elliptic problem at the critical exponent in pierced non-symmetric domains.* Comm. Partial Differential Equations **35** (2010), no. 8, 14191457.

[25] Y. Ge, R. Jing, F. Pacard: *Bubble towers for supercritical semilinear elliptic equations.* J. Funct. Anal. **221** (2005), no. 2, 251–302.

[26] M. Grossi, F. Takahashi: *Nonexistence of multi-bubble solutions to some elliptic equations on convex domains*, J. Funct. Anal. **259** (2010), 904–917.

[27] Z.C. Han: *Asymptotic approach to singular solutions for nonlinear elliptic equations involving critical Sobolev exponent*, Ann. Inst. H. Poincaré, Anal. Non Lin. **8** (1991), 159–174.

[28] G. Hardy, J. Littlewood: *Some properties of fractional integrals,* Math. Z. **27** (1928) 565–606.

[29] J. Kazdan, F. Warner: *Remarks on some quasilinear elliptic equations,* Comm. Pure Appl. Math. **28** (1975), 567–597.

[30] G. Li, S. Yan, J. Yang: *An elliptic problem with critical growth in domains with shrinking holes.* J. Differential Equations **198** (2004), no. 2, 275–300.

[31] A.M. Micheletti, A. Pistoia: *On the effect of the domain geometry on the existence of sign changing solutions to elliptic problems with critical and supercritical growth.* Nonlinearity **17** (2004), no. 3, 851–866.

[32] M. Musso, A. Pistoia: *Double blow-up solutions for a Brezis–Nirenberg type problem.* Commun. Contemp. Math. **5** (2003), no. 5, 775–802.

[33] M. Musso, A. Pistoia: *Sign changing solutions to a nonlinear elliptic problem involving the critical Sobolev exponent in pierced domains.* J. Math. Pures Appl. **86** (2006), no. 6, 510–528.

[34] M. Musso, A. Pistoia: *Sign changing solutions to a Bahri–Coron problem in pierced domains.* Discrete Contin. Dyn. Syst. **21** (2008), no. 1, 295–306.

[35] M. Musso, A. Pistoia: *Multispike solutions for a nonlinear elliptic problem involving the critical Sobolev exponent.* Indiana Univ. Math. J. **51** (2002), no. 3, 541–579,

[36] M. Musso, A. Pistoia: *Tower of bubbles for almost critical problems in general domains,* J. Math. Pures Appl. **93** (2010), 1–40.

[37] A. Pistoia, O. Rey: *Multiplicity of solutions to the supercritical Bahri–Coron's problem in pierced domains.* Adv. Differential Equations **11** (2006), no. 6, 647–666.

[38] A. Pistoia, T. Weth: *Sign changing bubble tower solutions in a slightly subcritical semilinear Dirichlet problem*, Ann. Inst. H. Poincaré Anal. Non Linéaire **24** (2007), 325–340.

[39] S.I. Pohožaev: *On the eigenfunctions of the equation* $\Delta u + \lambda f(u) = 0$, Soviet Math. Dokl. **6** (1965), 1408–1411.

[40] O. Rey: *The role of the Green function in a nonlinear elliptic equation involving the critical Sobolev exponent*, J. Funct. Anal. **89** (1990), 1–52.

[41] O. Rey: *Proof of two conjectures of H. Brezis and L.A. Peletier*, Manuscripta Math. **65** (1989), 19–37.

[42] O. Rey: *On a variational problem with lack of compactness: the effect of small holes in the domain* C. R. Acad. Sci. Paris Sér. I Math. **308** (1989), no. 12, 349–352.

[43] O. Rey: *Blow-up points of solutions to elliptic equations with limiting nonlinearity*, Differential Integral Equations **4** (1991), 1155–1167.

[44] G. Talenti: *Best constant in Sobolev inequality*, Ann. Mat. Pura Appl. **110** (1976), 353–372.

[45] S. Sobolev: *On a theorem of functional analysis,* AMS Transl. Series 2 **34** (1963) 39–68.

Angela Pistoia
Dipartimento di Metodi e Modelli Matematici
Università di Roma "La Sapienza"
via Antonio Scarpa 16
I-00161 Roma, Italy
e-mail: `pistoia@dmmm.uniroma1.it`

Concentration Analysis and Applications to PDE
Trends in Mathematics, 85–116

A General Theorem for the Construction of Blowing-up Solutions to Some Elliptic Nonlinear Equations via Lyapunov–Schmidt's Finite-dimensional Reduction

Frédéric Robert and Jérôme Vétois

Abstract. We prove a general finite-dimensional reduction theorem for critical equations of scalar curvature type. Solutions of these equations are constructed as a sum of peaks. The use of this theorem reduces the proof of existence of multi-peak solutions to some test-functions estimates and to the analysis of the interactions of peaks.

Mathematics Subject Classification (2010). 35J35, 35J60, 58J05, 35B44.

Keywords. Nonlinear elliptic equations, blow-up, conformal invariance.

1. Introduction and statement of the result

Let (M, g) be a compact Riemannian manifold of dimension $n \geq 3$ without boundary. We let $H_1^2(M)$ be the completion of $C^\infty(M)$ for the norm $\|\cdot\|_{H_1^2} := \|\cdot\|_2 + \|\nabla \cdot\|_2$. We let $h \in L^\infty(M)$ be such that the operator $\Delta_g + h$ is coercive, that is $\lambda_1(\Delta_g + h) > 0$, where $\Delta_g := -\mathrm{div}_g(\nabla)$ is the Laplace–Beltrami operator. Non-positive examples of such h's are after the theorem. We define $2^\star := \frac{2n}{n-2}$ and $H : \mathbb{R} \to \mathbb{R}$ such that $\{H(x) = |x| \text{ for all } x \in \mathbb{R}\}$ or $\{H(x) = x_+ := \max\{x, 0\} \text{ for all } x \in \mathbb{R}\}$. Given $f \in C^0(M)$, $q \in (2, 2^\star]$, and $G \in C^2(H_1^2(M))$, we give a general theorem to construct solutions $v \in H_1^2(M)$ to the equation

$$\Delta_g v + hv = fH(v)^{q-2}v + G'(v) \text{ in the distributional sense on } M \tag{1}$$

of the form

$$v = u_0 + \sum_{i=1}^{k} W_{\kappa_i,\delta_i,\xi_i} + \text{ remainder},$$

The authors are partially supported by the ANR grant ANR-08-BLAN-0335-01.

where $k \in \mathbb{N}$, $(\kappa_i)_{i=1,\dots,k} \in \{-1,+1\}$, $(\delta_i)_{i=1,\dots,k} \in (0,+\infty)$, $(\xi_i)_{i=1,\dots,k} \in M$ are the parameters, and the $W_{\kappa,\delta,\xi}$'s are peaks defined in (12) below and are C^1 with respect to the parameters. The function $u_0 \in H_1^2(M)$ is a distributional solution to

$$\Delta_g u_0 + h_0 u_0 = f_0 H(u_0)^{2^\star-2} u_0 + G_0'(u_0), \tag{2}$$

where $h_0 \in L^\infty(M)$ is such that $\lambda_1(\Delta_g + h_0) > 0$, $f_0 \in C^0(M)$, and $G_0 \in C^2(H_1^2(M))$ is of *subcritical type*, see Definition 2.1 below. Examples of nonlinearities of *subcritical type* are maps like $u \mapsto \int_M a(x)|u|^r\,dx$, where $a \in L^\infty(M)$ and $2 \le r < 2^\star$. Solutions to (1) and (2) are critical points respectively for the functionals

$$J(v) := \frac{1}{2}\int_M \left(|\nabla v|_g^2 + hv^2\right) dv_g - F(v),$$
$$J_0(v) := \frac{1}{2}\int_M \left(|\nabla v|_g^2 + h_0 v^2\right) dv_g - F_0(v),$$

where dv_g is the Riemannian element of volume, and

$$F(v) := \frac{1}{q}\int_M fH(v)^q\,dv_g + G(v) \quad \text{and} \quad F_0(v) := \frac{1}{2^\star}\int_M f_0 H(v)^{2^\star}\,dv_g + G_0(v)$$

for all $v \in H_1^2(M)$. We introduce the kernel of the linearization of (2) by

$$K_0 := \left\{\varphi \in H_1^2(M) /\, \Delta_g\varphi + h_0\varphi = F_0''(u_0)\varphi\right\}. \tag{3}$$

Since the operator $\varphi \mapsto (\Delta_g + h_0)^{-1}(F_0''(u_0)\varphi)$ is compact on $H_1^2(M)$, we get that $d := \dim_{\mathbb{R}} K_0 < +\infty$. We let $u \in C^1(B_1(0) \subset \mathbb{R}^d, H_1^2(M))$ be such that $u(0) = u_0$, and we assume that

$$K_0 = \mathrm{Span}\{\Pi_{K_0}^{h_0}(\partial_{z_i} u(0)) /\, i = 1,\dots,d\}, \tag{4}$$

where $\Pi_{K_0}^{h_0}$ is the orthogonal projection on K_0 with respect to the scalar product $(u,v) \mapsto (u,v)_{h_0} := \int_M ((\nabla u, \nabla v)_g + h_0 uv)\,dv_g$. We consider a finite covering $(U_\gamma)_{\gamma\in\mathcal{C}}$ of M of *parallel type* (see Definition 2.2), and we choose a correspondance $i \mapsto \gamma_i \in \mathcal{C}$ for all $i \in \{1,\dots,k\}$. For any $\varepsilon > 0$, $N > 0$, and $k \in \mathbb{N}$, we define

$$\mathcal{D}_k(\varepsilon, N) := \Big\{((\delta_i)_i, (\xi_i)_i) \in (0,\varepsilon)^k \times M^k /\, \xi_i \in U_{\gamma_i},\, |\delta_i^{2^\star-q} - 1| < \varepsilon,$$
$$\text{and } \frac{\delta_i}{\delta_j} + \frac{\delta_j}{\delta_i} + \frac{d_g(\xi_i,\xi_j)^2}{\delta_i\delta_j} > N \text{ for all } i \ne j \in \{1,\dots,k\}\Big\}.$$

We define the error term

$$\begin{aligned} R(z,(\delta_i)_i,(\xi_i)_i) := \Big\| u(z) + \sum\nolimits_{i=1}^k W_{\kappa_i,\delta_i,\xi_i} \\ - (\Delta_g + h)^{-1}\Big(F'\Big(u(z) + \sum\nolimits_{i=1}^k W_{\kappa_i,\delta_i,\xi_i}\Big)\Big)\Big\|_{H_1^2}. \end{aligned} \tag{5}$$

Theorem 1.1. *We fix $k \in \mathbb{N}$, $\nu_0, C_0 > 0$, $\theta \in (0,1)$, $h_0 \in L^\infty(M)$ such that $\lambda_1(\Delta_g + h_0) > 0$, $f_0 \in C^0(M)$, $u_0 \in H_1^2(M)$, and $G_0 \in C^{2,\theta}_{\mathrm{loc}}(H_1^2(M))$ of subcritical type. We define K_0 as in (3), we let d be its dimension and β_0 be a basis of K_0.*

We fix $(\kappa_i)_i \in \{-1,+1\}^k$. *Then there exist* $N > 0$ *and* $\varepsilon > 0$ *such that for any* $q \in (2, 2^\star]$, $h \in L^\infty(M)$, $f \in C^0(M)$, $G \in C^{2,\theta}_{\text{loc}}(H^2_1(M))$, *and* $u \in C^1(B_1(0), H^2_1(M))$ *such that*

$$u(0) = u_0\,,\ \|u\|_{C^1(B_1(0),H_1^2)} \le C_0\,,\ f_0(\xi_i) \ge \nu_0 \text{ for all } i = 1, \dots, k\,, \tag{6}$$

$$\|h - h_0\|_\infty + \|f - f_0\|_{C^0(M)} + d_{C^{2,\theta}_B}(G, G_0) + (2^\star - q) < \varepsilon\,, \tag{7}$$

(*see Definition* 2.3 *for the distance* $d_{C^{2,\theta}_B}$) *and for any* $z \in B_1(0)$, *if*

$$\left|\det_{\beta_0}(\Pi^{h_0}_{K_0}(\partial_{z_1} u(z)), \dots, \Pi^{h_0}_{K_0}(\partial_{z_d} u(z)))\right| \ge \nu_0 \prod_{i=1}^d \|\partial_{z_i} u(z)\|_{H_1^2} > 0\,, \tag{8}$$

then there exists $\phi \in C^1(B_\varepsilon(0) \times \mathcal{D}_k(\varepsilon, N), H^2_1(M))$ *such that* $u(z, (\delta_i)_i, (\xi_i)_i) := u(z) + \sum_{i=1}^k W_{\kappa_i,\delta_i,\xi_i} + \phi(z, (\delta_i)_i, (\xi_i)_i)$ *is a critical point of* J *if and only if* $(z, (\delta_i)_i, (\xi_i)_i)$ *is a critical point of* $(z, (\delta_i)_i, (\xi_i)_i) \mapsto J(u(z, (\delta_i)_i, (\xi_i)_i))$ *in* $B_\varepsilon(0) \times \mathcal{D}_k(\varepsilon, N)$. *Moreover, we have that*

$$\|\phi(z, (\delta_i)_i, (\xi_i)_i)\|_{H_1^2} \le C \cdot R(z, (\delta_i)_i, (\xi_i)_i)\,,$$

where C *is a constant depending on* (M, g), k, ν_0, θ, C_0, u_0, h_0, f_0, *and* G_0.

Miscellaneous remarks

1. The implicit definition of $\phi(z, (\delta_i)_i, (\xi_i)_i)$ is in (61) of Proposition 5.1.
2. In addition to Theorem 1.1, we have that

$$\left|J(u(z, (\delta_i)_i, (\xi_i))) - J\Big(u(z) + \sum_{i=1}^k W_{\kappa_i,\delta_i,\xi_i}\Big)\right| \le C \cdot R(z, (\delta_i)_i, (\xi_i)_i)^2$$

   for all $(z, (\delta_i)_i, (\xi_i)_i)) \in B_\varepsilon(0) \times \mathcal{D}_k(\varepsilon, N)$, where C is a constant depending on (M, g), k, ν_0, C_0, u_0, G_0.
3. Theorem 1.1 is valid under a little more general hypothesis on G. There exists $\tilde{R} > 0$ depending only on (M, g), u_0, and k such that the same conclusion of the theorem holds if G_0 and G satisfy the following

$$G_0 \in C^2(B_{\tilde{R}}(0)),\ G \in C^{2,\theta}(B_{\tilde{R}}(0)),$$
$$\|G\|_{C^{2,\theta}}(B_{\tilde{R}}(0)) \le C_0\,, \quad \text{and} \quad \|G - G_0\|_{C^2(B_{\tilde{R}}(0))} < \varepsilon\,.$$

4. We have assumed for convenience a L^∞-control of the potentials h_0 and h. If they are only controled in $L^{n/2}$, it suffices to include them in the perturbations G_0 and G.
5. As one checks, if $h \ge 0$ and $h \not\equiv 0$, one gets $\lambda_1(\Delta_g + h) > 0$. As a consequence, $\lambda_1(\Delta_g + h - \alpha) > 0$ for such h and all $\alpha < \lambda_1(\Delta_g + h)$.

As a consequence of Theorem 1.1, finding solutions to (1) reduces to computing the expansion of $J(u(z, (\delta_i)_i, (\xi_i)))$ and controling the rest $R(z, (\delta_i)_i, (\xi_i)_i)$. In particular, Theorem 1.1 covers the general reduction theory in the recent articles Esposito–Pistoia–Vétois [6], Micheletti–Pistoia–Vétois [9], Pistoia–Vétois [10], and Robert–Vétois [12]. This finite-reduction method is very classical and has proved to be very powerful in the last decades to find blowing-up solutions to critical equations. The litterature on this issue is abundant: here, we refer to the early reference Rey [11], and to Brendle [3], Brendle–Marques [4], del Pino–Musso–Pacard–Pistoia [5], and Guo–Li–Wei [8] for more recent references. The list of constributions above does not pretend to exhaustivity: we refer to the references of the above papers and also to the monograph [1] by Ambrosetti–Malchiodi for further bibliographic complements. A general reference on Lyapunov–Schmidt's reduction, including the group action point of view, is the monograph [7] by Falaleev–Loginov–Sidorov–Sinitsyn.

2. Definitions and notations

2.1. Nonlinearities of subcritical type

Definition 2.1. Let $G_0 \in C^2(H_1^2(M))$. We say that G_0 is of *subcritical type* if for all sequences $(u_p)_p, (v_p)_p, (w_p)_p \in H_1^2(M)$ converging weakly respectively to $u, v, w \in H_1^2(M)$, we have that

$$G_0(u_p) \to G_0(u)\,,\; G_0'(u_p)(v_p) \to G_0'(u)(v),$$
$$\text{and}\;\; G_0''(u_p)(v_p, w_p) \to G_0''(u)(v, w)$$

when $p \to +\infty$.

2.2. Covering of parallel type

Definition 2.2. We say that $(U_\gamma)_{\gamma\in\mathcal{C}}$ is a covering of parallel type if $\cup_\gamma U_\gamma = M$ and for any $\gamma \in \mathcal{C}$, U_γ is open and there exists n smooth vector fields $e_i^{(\gamma)} : U_\gamma \to TM$ such that for any $\xi \in U_\gamma$, $\{e_1^{(\gamma)}(\xi), \dots, e_n^{(\gamma)}(\xi)\}$ is an orthonormal basis of $T_\xi M$, the tangent space at the point ξ.

Since (M, g) is compact, it follows from the Gram–Schmidt orthogonalisation procedure that a finite covering of *parallel type* always exists. A manifold is parallelizable if there exists a smooth global orthonormal basis.

In the sequel, we let $(U_\gamma)_{\gamma\in\mathcal{C}}$ be a fixed finite covering of parallel type of M. With a slight abuse of notation, for any $\gamma \in \mathcal{C}$, and any $\xi \in U_\gamma$, we define $e_j(\xi) = e_j^{(\gamma)}(\xi)$ for $j = 1, \dots, n$, where $e_j^{(\gamma)}$ is as in Definition 2.2. In other words, for any $\gamma \in \mathcal{C}$, there exists n smooth maps $e_1, \dots, e_j : U_\gamma \to TM$ such that for any $\xi \in U_\gamma$, $(e_1(\xi), \dots, e_n(\xi))$ is an orthonormal basis of $T_\xi M$. We can then assimilate smoothly the tangent space $T_\xi M$ at $\xi \in U_\gamma$ to $\mathbb{R}^n$ via the map

$$\begin{array}{rccc} \Phi_\xi : & \mathbb{R}^n & \to & T_\xi M \\ & X & \mapsto & \sum_{j=1}^n X^j e_j(\xi). \end{array} \tag{9}$$

2.3. The distance on $C_B^{2,\theta}$

Definition 2.3. Let E be a Banach space. We define $C_B^{2,\theta}(E)$ as the set of functions that are in $C^{2,\theta}(B)$ for any bounded open set $B \subset E$: we endow $C_B^{2,\theta}(E)$ with the topology inherited from the natural associated family of semi-norms. This topology is metrizable with the distance

$$d_{C_B^{2,\theta}}(G_1, G_2) := \sup_{p\in\mathbb{N}} \frac{\|G_1 - G_2\|_{C^{2,\theta}(B_p(0))}}{2^p(1 + \|G_1 - G_2\|_{C^{2,\theta}(B_p(0))})} \quad \text{for all } G_1, G_2 \in C_B^{2,\theta}(E).$$

2.4. The peaks $W_{\kappa,\delta,\xi}$

We consider a function $\Lambda \in C^\infty(M \times M)$ such that, defining $\Lambda_\xi := \Lambda(\xi, \cdot)$ for all $\xi \in M$, we have that

$$\Lambda_\xi > 0 \text{ and } \Lambda_\xi(\xi) = 1 \text{ for all } \xi \in M. \tag{10}$$

We then define a metric $g_\xi := \Lambda_\xi^{\frac{4}{n-2}} g$ for all $\xi \in M$ conformal to g. Since Λ is continuous, there exists $C > 0$ such that

$$\frac{1}{C} g \le g_\xi \le Cg \tag{11}$$

for all $\xi \in M$. The compactness of M yields the existence of $r_0 > 0$ such that the injectivity radius of the metric g_ξ satisfies $i_{g_\xi}(M) \ge r_0$ for all $\xi \in M$. We let $\chi \in C^\infty(\mathbb{R})$ be such that $\chi(t) = 1$ for $t \le r_0/3$, $\chi(t) = 0$ for all $t \ge r_0/2$ and $0 \le \chi \le 1$.

For $\kappa \in \{-1, +1\}$, $\delta > 0$, and $\xi \in M$ such that $f_0(\xi) > 0$, a bubble is defined as

$$W_{\kappa,\delta,\xi}(x) := \kappa\chi(d_{g_\xi}(x,\xi))\Lambda_\xi(x)\left(\frac{\delta\sqrt{\frac{n(n-2)}{f_0(\xi)}}}{\delta^2 + d_{g_\xi}(x,\xi)^2}\right)^{\frac{n-2}{2}} + B_{\delta,\xi}(x) \tag{12}$$

for all $x \in M$, where $(\delta, \xi) \mapsto B_{\delta,\xi}$ is C^1 from $(0, +\infty) \times M$ to $H_1^2(M)$ and

$$\|B_{\delta,\xi}\|_{H_1^2} + \delta\|\partial_\delta B_{\delta,\xi}\|_{H_1^2} + \delta\|\nabla_\xi B_{\delta,\xi}\|_{H_1^2} \le \epsilon(\delta) \tag{13}$$

for all $\delta > 0$ and $\xi \in M$, where $\lim_{\delta\to 0}\epsilon(\delta) = 0$. If $H = (\cdot)_+$, we require that $\kappa = 1$.

2.5. Derivatives of the peaks

We let $D_1^2(\mathbb{R}^n)$ be the completion of $C_c^\infty(\mathbb{R}^n)$ for the norm $u \mapsto \|\nabla u\|_2$. Given $a > 0$, we are interested in solutions $U \in D_1^2(\mathbb{R}^n)$ to the equation

$$\Delta_{\text{Eucl}} U_a = a U_a^{2^\star - 1} \text{ in } \mathbb{R}^n, \tag{14}$$

where Eucl is the Euclidean metric on $\mathbb{R}^n$. As one checks, the Lie group $(0, +\infty) \times \mathbb{R}^n$ (with the relevant structure) leaves the solution to (14) invariant via the action

$$(\delta, x_0) \in (0, +\infty) \times \mathbb{R}^n \mapsto \delta^{-\frac{n-2}{2}} U_a(\delta^{-1}(\cdot - x_0)). \tag{15}$$

For $a > 0$, we define

$$U_a(x) := \left(\frac{\sqrt{\frac{n(n-2)}{a}}}{1 + |x|^2}\right)^{\frac{n-2}{2}} \quad \text{for all } x \in \mathbb{R}^n.$$

As easily checked, we have that $U_a \in D_1^2(\mathbb{R}^n)$ is a solution to (14). Therefore, the action of the Lie algebra of $(0,+\infty)\times\mathbb{R}^n$ yields elements of the set K_{BE} of solutions $V \in D_1^2(\mathbb{R}^n)$ of the linearized equation

$$\Delta_{\mathrm{Eucl}} V = (2^\star - 1) a U_a^{2^\star - 2} V \text{ in } \mathbb{R}^n. \tag{16}$$

Conversely, it follows from Bianchi–Egnell [2] that this actions is onto, that is

$$K_{BE} = \mathrm{Span}\{V_j / \, j = 0, \dots, n\},$$

where

$$V_0 := \frac{2}{n-2}\left(\frac{a}{n(n-2)}\right)^{(n-2)/4} \frac{\partial}{\partial\delta}(\delta^{-\frac{n-2}{2}} U_a(\delta^{-1}\cdot))_{\delta=1} = \frac{|x|^2 - 1}{(1+|x|^2)^{\frac{n}{2}}},$$
$$V_j := \frac{-1}{n-2}\left(\frac{a}{n(n-2)}\right)^{(n-2)/4} \partial_{x_j} U_a = \frac{x_j}{(1+|x|^2)^{\frac{n}{2}}} \text{ for all } j = 1, \dots, n.$$

The functions V_j form an orthonormal basis of K_{BE} for the scalar product $(u,v) \mapsto \int_{\mathbb{R}^n} (\nabla u, \nabla v)\, dx$. Rescaling and pulling-back on M, for any $\delta > 0$, $\xi \in M$, and $X \in T_\xi M$, we define

$$Z_{\delta,\xi}(x) := \chi(d_{g_\xi}(x,\xi))\Lambda_\xi(x)\delta^{\frac{n-2}{2}} \frac{d_{g_\xi}(x,\xi)^2 - \delta^2}{(\delta^2 + d_{g_\xi}(x,\xi)^2)^{\frac{n}{2}}}, \tag{17}$$

$$Z_{\delta,\xi,X}(x) := \chi(d_{g_\xi}(x,\xi))\Lambda_\xi(x)\delta^{\frac{n}{2}} \frac{\langle (\exp_\xi^{g_\xi})^{-1}(x), X\rangle_{g_\xi(\xi)}}{(\delta^2 + d_{g_\xi}(x,\xi)^2)^{\frac{n}{2}}} \tag{18}$$

for all $x \in M$. We let $(U_\gamma)_\gamma$ be as in Definition 2.2. Here and in the sequel, $\exp_\xi^{g_\xi}$ denotes the exponential map at the point $\xi \in M$ with respect to the metric g_ξ. For $\gamma \in \mathcal{C}$, $\xi \in U_\gamma$, and $\delta > 0$, we define

$$Z_{\delta,\xi,j} := Z_{\delta,\xi,e_j(\xi_i)} \text{ for } j = 1, \dots, n \text{ and } Z_{\delta,\xi,0} := Z_{\delta,\xi}, \tag{19}$$

where the $e_j(\xi)$'s are defined in Definition 2.2: we have omitted the index γ for clearness. Since the isometric assimilation (9) of the tangent space to $\mathbb{R}^n$ is smooth with respect to $\xi \in U_\gamma$, we define

$$\begin{array}{rccc} \tilde{\exp}_\xi^{g_\xi}: & \mathbb{R}^n & \to & M \\ & X & \mapsto & \exp_\xi^{g_\xi}(\Phi_\xi(X)). \end{array} \tag{20}$$

2.6. Sobolev inequalities

It follows from Sobolev's Theorem that $D_1^2(\mathbb{R}^n)$ is embedded continuously in $L^{2^\star}(\mathbb{R}^n)$ and that for any $\varphi \in D_1^2(\mathbb{R}^n)$, we have that

$$\|\varphi\|_{2^\star} \le K(n,2)\|\nabla\varphi\|_2 \text{ with } K(n,2) := 2(n\,(n-2)\,\omega_n^{2/n})^{-1/2}. \tag{21}$$

On the compact manifold (M,g), $H_1^2(M)$ is embedded in $L^{2^\star}(M)$, and there exists $A > 0$ such that for any $\phi \in H_1^2(M)$, we have that

$$\|\phi\|_{2^\star} \le A\|\phi\|_{H_1^2}. \tag{22}$$

2.7. Riesz correspondence

We let $\epsilon_0 > 0$ be such that for $h \in L^\infty(M)$ such that $\|h - h_0\|_\infty < \varepsilon_0$, we have that $\|h\|_\infty \leq \|h_0\|_\infty + 1$ and $\lambda_1(\Delta_g + h) \geq \lambda_1(\Delta_g + h_0)/2 > 0$. With a slight abuse of notation, we define

$$\left\{ \begin{array}{llll} \Delta_g + h: & H_1^2(M) & \to & (H_1^2(M))' \\ & u & \mapsto & (v \mapsto (u,v)_h := \int_M ((\nabla u, \nabla v)_g + huv)\, dv_g) \end{array} \right\},$$

and its inverse is denoted as $(\Delta_g + h)^{-1}$. For $\tau \in L^{\frac{2n}{n+2}}(M)$, it follows from Sobolev's Theorem (see (22)) that the map $T_\tau : v \mapsto \int_M \tau v\, dv_g$ is defined and continuous for $v \in H_1^2(M)$: we will then write $(\Delta_g + h)^{-1}(\tau) := (\Delta_g + h)^{-1}(T_\tau)$. It then follows from regularity theory that for $\|h - h_0\| < \varepsilon_0$, we have that

$$\|(\Delta_g + h)^{-1}(\tau)\|_{H_1^2} \leq C(h_0, \varepsilon_0)\|\tau\|_{\frac{2n}{n+2}}, \tag{23}$$

where $C(h_0, \varepsilon_0) > 0$ depends only on (M, g), $h_0 \in L^\infty(M)$, and $\varepsilon_0 > 0$.

2.8. Notation

In the sequel, $C, C_1, C_2, \ldots$ will denote positive constants depending only on (M, g), k, ν_0, θ, C_0, u_0, h_0, f_0, and G_0. We will often use the same notation C or C_i $(i \geq 1)$ for different constants from line to line, and even in the same line.

The notation $\omega_{a,b,\ldots}(x)$ will denote a constant depending on $a, b, \ldots, x$, (M, g), k, ν_0, θ, C_0, u_0, h_0, f_0, and G_0 and such that $\lim_{x \to l} \omega_{a,b,\ldots}(x) = 0$, where $l \in \{0, +\infty\}$ will be explicit for each statement.

3. Preliminary computations 1: Rescaling and pull-back

The objective of this section and the following is to express qualitatively the transfer of the action of $(0, +\infty) \times \mathbb{R}^n$ on $D_1^2(\mathbb{R}^n)$ to an infinitesimal action on $H_1^2(M)$. We fix $\gamma \in \mathcal{C}$, where $\mathcal{C}$ is as in Definition 2.2. We choose a function $F \in C^\infty(M \times M)$ such that $F(\xi, x) = 0$ if $d_{g_\xi}(\xi, x) \geq r_0$ for $\xi, x \in M$. For $\varphi \in D_1^2(\mathbb{R}^n)$, we define for $\xi \in U_\gamma$ and $\delta > 0$

$$\mathrm{Resc}^F_{\delta,\xi}(\varphi)(x) := F(\xi, x)\delta^{-\frac{n-2}{2}}\varphi(\delta^{-1}(\tilde{\exp}^{g_\xi}_\xi)^{-1}(x)) \tag{24}$$

for all $x \in M$. This transformation is the infinitesimal transfer via the exponential map of the action $(0, +\infty) \times \mathbb{R}^n$ on $D_1^2(\mathbb{R}^n)$ defined in (15). As a preliminary remark, it follows from (19) that

$$Z_{\delta_i,\xi_i,j} = \mathrm{Resc}^{F^{(1)}}_{\delta_i,\xi_i}(V_j) \text{ and } W_{\kappa_i,\xi_i,\delta_i} = \mathrm{Resc}^{F^{(2)}}_{\delta_i,\xi_i}(U_1) + B_{\delta_i,\xi_i} \tag{25}$$

with

$$F^{(1)}(\xi, x) := \chi(d_{g_\xi}(x, \xi))\Lambda_\xi(x),$$
$$F^{(2)}(\xi, x) := \kappa_i f_0(\xi)^{-(n-2)/4}\chi(d_{g_\xi}(x, \xi))\Lambda_\xi(x)$$

for all $\xi, x \in M$.

Proposition 3.1. *For all* $\varphi \in D_1^2(\mathbb{R}^n)$, $\delta > 0$, *and* $\xi \in U_\gamma$, *we have that* $\mathrm{Resc}_{\delta,\xi}(\varphi) \in H_1^2(M)$ *and*

$$\| \mathrm{Resc}^F_{\delta,\xi}(\varphi)\|_{H_1^2} \leq C_1(F)\|\varphi\|_{D_1^2}\,, \tag{26}$$

where $C_1(F) > 0$ *is independent of* $\xi \in U_\gamma$, $\delta > 0$, *and* $\varphi \in D_1^2(\mathbb{R}^n)$. *Moreover, for all* $\varphi, \psi \in D_1^2(\mathbb{R}^n)$, *and for all* $\delta, R > 0$ *and* $\xi \in U_\gamma$, *we have that*

$$\begin{aligned} \Bigg| \int_{B^{g_\xi}_{R\delta}(\xi)} & \left(\nabla \mathrm{Resc}^F_{\delta,\xi}(\varphi), \nabla \mathrm{Resc}^F_{\delta,\xi}(\psi)\right)_{g_\xi} dv_{g_\xi} \\ & - F(\xi,\xi)^2 \int_{\mathbb{R}^n} (\nabla\varphi, \nabla\psi)_{\mathrm{Eucl}}\, dx \Bigg| \leq \omega_{1,F,\varphi,\psi}(R) + \omega_{2,F,\varphi,\psi}(\delta), \end{aligned} \tag{27}$$

$$\int_{M\setminus B^{g_\xi}_{R\delta}(\xi)} |\nabla \mathrm{Resc}^F_{\delta,\xi}(\varphi)|^2_{g_\xi}\, dv_{g_\xi} \leq \omega_{1,F,\varphi,\psi}(R) + \omega_{2,\varphi,\psi}(\delta), \tag{28}$$

$$\int_M | \mathrm{Resc}^F_{\delta,\xi}(\varphi)|^2\, dv_{g_\xi} \leq \omega_{3,F,\varphi}(\delta), \tag{29}$$

where

$$\lim_{R\to+\infty} \omega_{1,F,\varphi,\psi}(R) = \lim_{\delta\to 0} \omega_{2,F,\varphi,\psi}(\delta) = \lim_{\delta\to 0} \omega_{3,F,\varphi}(\delta) = 0.$$

Proof of Proposition 3.1. We fix $\varphi, \psi \in D_1^2(\mathbb{R}^n)$. We consider a domain $D \subset M$. A change of variable yields

$$\int_D \left(\nabla \mathrm{Resc}^F_{\delta,\xi}(\varphi), \nabla \mathrm{Resc}^F_{\delta,\xi}(\psi)\right)_{g_\xi} dv_{g_\xi} = \int_{D_{\delta,\xi}} (\nabla(\phi_{\delta,\xi}\varphi), \nabla(\phi_{\delta,\xi}\psi))_{g_{\delta,\xi}}\, dv_{g_{\delta,\xi}}\,, \tag{30}$$

where $D_{\delta,\xi} := \delta^{-1}(\tilde{\exp}^{g_\xi}_\xi)^{-1}(D \cap B^{g_\xi}_{r_0}(\xi))$, $g_{\delta,\xi}(x) := ((\tilde{\exp}^{g_\xi}_\xi)^\star g)(\delta x)$, and

$$\phi_{\delta,\xi}(x) := F(\xi, \tilde{\exp}^{g_\xi}_\xi(\delta x))$$

for all $x \in \mathbb{R}^n$. Integrating (30) by parts yields

$$\begin{aligned} \int_D & \left(\nabla \mathrm{Resc}^F_{\delta,\xi}(\varphi), \nabla \mathrm{Resc}^F_{\delta,\xi}(\psi)\right)_{g_\xi} dv_{g_\xi} \\ & = \int_{D_{\delta,\xi}} \left(\phi^2_{\delta,\xi}(\nabla\varphi,\nabla\psi)_{g_{\delta,\xi}} + \phi_{\delta,\xi}(\Delta_{g_{\delta,\xi}}\phi_{\delta,\xi})\varphi\psi\right) dv_{g_{\delta,\xi}}\,. \end{aligned} \tag{31}$$

Since F is smooth, there exists $C(F) > 0$ such that

$$\begin{cases} |\phi_{\delta,\xi}(x) - \phi_{\delta,\xi}(0)| \leq C(F)\delta|x|\,, \ |\phi_{\delta,\xi}\Delta_{g_{\delta,\xi}}\phi_{\delta,\xi}(x)| \leq C(F)\delta^2, \\ |g_{\delta,\xi}(x) - g_{\delta,\xi}(0)| \leq C(F)\delta\,\mathrm{Eucl}\,, \text{ and } |dv_{g_{\delta,\xi}}(x) - dx| \leq C(F)\delta\, dx \end{cases} \tag{32}$$

for all $x \in B_{r_0/\delta}(0) \subset \mathbb{R}^n$. Since $\phi_{\delta,\xi}(0) = F(\xi,\xi)$ and $g_{\delta,\xi}(0) = \text{Eucl}$ the Euclidean metric in $\mathbb{R}^n$, plugging (32) into (31) yields

$$\begin{aligned} &\left| \int_D \left(\nabla \operatorname{Resc}^F_{\delta,\xi}(\varphi), \nabla \operatorname{Resc}^F_{\delta,\xi}(\psi)\right)_{g_\xi} dv_{g_\xi} - F(\xi,\xi)^2 \int_{D_{\delta,\xi}} (\nabla\varphi, \nabla\psi)_{\text{Eucl}}\, dx \right| \\ &\leq C(F) \int_{D_{\delta,\xi}} (\delta|\nabla\varphi|\cdot|\nabla\psi| + \delta^2|\varphi|\cdot|\psi|)\, dx \\ &\leq C(F)\delta\|\nabla\varphi\|_2\|\nabla\psi\|_2 + C(F)\sqrt{\delta^2\int_{B_{r_0/\delta}(0)} \varphi^2\, dx} \cdot \sqrt{\delta^2\int_{B_{r_0/\delta}(0)} \psi^2\, dx}\,. \end{aligned} \tag{33}$$

Independently, for any $R > 0$, we have that

$$\begin{aligned} \delta^2 \int_{B_{r_0/\delta}(0)} \varphi^2\, dx &\leq \delta^2 \int_{B_{r_0/\delta}(0)\setminus B_R(0)} \varphi^2\, dx + \delta^2 \int_{B_R(0)} \varphi^2\, dx \qquad (34) \\ &\leq \delta^2 \cdot \left(\int_{B_{r_0/\delta}(0)\setminus B_R(0)} dx\right)^{\frac{2}{n}} \left(\int_{B_{r_0/\delta}(0)\setminus B_R(0)} |\varphi|^{2^\star}\, dx\right)^{\frac{2}{2^\star}} \\ &\quad + \delta^2 \cdot \left(\int_{B_R(0)} dx\right)^{\frac{2}{n}} \left(\int_{B_R(0)} |\varphi|^{2^\star}\, dx\right)^{\frac{2}{2^\star}} \\ &\leq C r_0^2 \left(\int_{\mathbb{R}^n\setminus B_R(0)} |\varphi|^{2^\star}\, dx\right)^{\frac{2}{2^\star}} + C\delta^2 R^2 \left(\int_{\mathbb{R}^n} |\varphi|^{2^\star}\, dx\right)^{\frac{2}{2^\star}}. \end{aligned}$$

Since $\varphi \in D_1^2(\mathbb{R}^n)$, it follows from Sobolev's inequality (21) that $\varphi \in L^{2^\star}(\mathbb{R}^n)$ and

$$\lim_{\delta\to 0} \delta^2 \int_{B_{r_0/\delta}(0)} \varphi^2\, dx = 0\,. \tag{35}$$

As a consequence, for all $\xi \in U_\gamma$, all $\delta > 0$, and all domain $D \subset M$, we have that

$$\begin{aligned} &\left| \int_D \left(\nabla \operatorname{Resc}^F_{\delta,\xi}(\varphi), \nabla \operatorname{Resc}^F_{\delta,\xi}(\psi)\right)_{g_\xi} dv_{g_\xi} - F(\xi,\xi)^2 \int_{D_{\delta,\xi}} (\nabla\varphi, \nabla\psi)_{\text{Eucl}}\, dx \right| \\ &\qquad \leq \omega_{4,F,\varphi,\psi}(\delta), \end{aligned} \tag{36}$$

where $\lim_{\delta\to 0}\omega_{4,F,\varphi,\psi}(\delta) = 0$. Taking alernatively $D := B^{g_\xi}_{R\delta}(\xi)$ or $D := M \setminus B^{g_\xi}_{R\delta}(\xi)$, and letting $R \to +\infty$ yields (27) and (28). Taking $R = 0$ in (34), taking $D := M$ and $\psi = \varphi$ in (33), and using Sobolev's inequality (21), we get that

$$\int_M \left|\nabla \operatorname{Resc}^F_{\delta,\xi}(\varphi)\right|^2_{g_\xi} dv_{g_\xi} \leq C(F)\int_{\mathbb{R}^n} |\nabla\varphi|^2\, dx + C(F)\left(\int_{\mathbb{R}^n} |\varphi|^{2^\star}\, dx\right)^{\frac{2}{2^\star}} \tag{37}$$

for $\delta < 1$. A change of variable and Hölder's inequality yields

$$\int_M \operatorname{Resc}_{\delta,\xi}(\varphi)^2\, dv_{g_\xi} = \delta^2 \int_{B_{r_0/\delta}(0)} |\phi_{\delta,\xi}\varphi|^2\, dv_{g_{\delta,\xi}} \leq C(F)\delta^2 \int_{B_{r_0/\delta}(0)} \varphi^2\, dx\,.$$

inequality (11), (35), and the latest inequality. Assertion (26) follows from (37), inequality (11), Sobolev's inequality (21) and (29). □

As a consequence, we get the following orthogonality property:

Proposition 3.2. *Let $\varphi, \psi \in D_1^2(\mathbb{R}^n)$ be two functions and $h \in L^\infty(M)$ such that $\|h\|_\infty < C_1$. Then for any $((\delta_i)_i, (\xi_i)_i) \in \mathcal{D}_k(\varepsilon, N)$, we have that*

$$\left|(\mathrm{Resc}^F_{\delta_i,\xi_i}(\varphi), \mathrm{Resc}^F_{\delta_j,\xi_j}(\psi))_h - \delta_{i,j} F(\xi,\xi)\int_{\mathbb{R}^n}(\nabla\varphi, \nabla\psi)\,dx\right| \le \omega_{5,F,C_1,\varphi,\psi}(\varepsilon, N)$$

for all $i, j \in \{1, \dots, k\}$, where $\lim_{\varepsilon\to 0\,;\,N\to+\infty}\omega_{5,F,C_1,\varphi,\psi}(\varepsilon,N) = 0$. Here, $\delta_{i,j} = 1$ if $i = j$ and 0 otherwise.

Proof of Proposition 3.2. We let $R > 0$ be a positive number. We have that

$$\begin{aligned}&\left|\int_M (\nabla\,\mathrm{Resc}^F_{\delta_i,\xi_i}(\varphi), \nabla\,\mathrm{Resc}^F_{\delta_j,\xi_j}(\psi))_g\,dv_g\right|\\ &\quad\le \int_{M\setminus B^g_{R\delta_i}(\xi_i)}\cdots + \int_{M\setminus B^g_{R\delta_j}(\xi_j)}\cdots + \int_{B^g_{R\delta_i}(\xi_i)\cap B^g_{R\delta_j}(\xi_j)}\cdots\end{aligned} \tag{38}$$

It follows from (11), and assertions (28) and (26) of Proposition 3.1 that

$$\begin{aligned}&\int_{M\setminus B^g_{R\delta_i}(\xi)}\left|(\nabla\,\mathrm{Resc}^F_{\delta_i,\xi_i}(\varphi), \nabla\,\mathrm{Resc}^F_{\delta_j,\xi_j}(\psi))_g\right|dv_g\\ &\le \sqrt{\int_{M\setminus B^{g_\xi}_{R\delta_i}(\xi)}|\nabla\,\mathrm{Resc}^F_{\delta_i,\xi_i}(\varphi)|^2_{g_\xi}\,dv_{g_\xi}}\cdot\|\mathrm{Resc}^F_{\delta_j,\xi_j}(\psi)\|_{H_1^2} \le \omega_{6,F,\varphi,\psi}(R),\end{aligned} \tag{39}$$

where $\lim_{R\to+\infty}\omega_{6,F,\varphi,\psi}(R) = 0$.

We first assume that $i \ne j$. If $B^g_{R\delta_i}(\xi_i) \cap B^g_{R\delta_j}(\xi_j) = \emptyset$, we get (42) from (38) and (39). We assume that $B^g_{R\delta_i}(\xi_i) \cap B^g_{R\delta_j}(\xi_j) \ne \emptyset$ and $i \ne j$. Then we have that $d_g(\xi_i, \xi_j) < R(\delta_i + \delta_j)$. Since $((\delta_i)_i, (\xi_i)_i) \in \mathcal{D}_k(\varepsilon, N)$, exchanging i and j if necessary, we then get that for N large enough and ε small enough that

$$\frac{\delta_i}{\delta_j} < \frac{2(1+R^2)}{N}\,. \tag{40}$$

Therefore, using (26), we get that

$$\begin{aligned}&\int_{B^g_{R\delta_i}(\xi_i)\cap B^g_{R\delta_j}(\xi_j)}\left|(\nabla\,\mathrm{Resc}^F_{\delta_i,\xi_i}(\varphi), \nabla\,\mathrm{Resc}^F_{\delta_j,\xi_j}(\psi))_g\right|dv_g\\ &\quad\le \|\mathrm{Resc}^F_{\delta_i,\xi_i}(\varphi)\|_{H_1^2(M)}\cdot\sqrt{\int_{B^{g_{\xi_j}}_{R\delta_i}(\xi_i)\cap B^{g_{\xi_j}}_{R\delta_j}(\xi_j)}|\nabla\,\mathrm{Resc}^F_{\delta_j,\xi_j}(\psi)|_{g_{\xi_j}}\,dv_{\xi_j}}\\ &\quad\le C(F)\|\varphi\|_{D_1^2}\cdot\sqrt{\int_{\delta_j^{-1}\tilde{\exp}_{\xi_j}^{-1}(B^{g_{\xi_j}}_{R\delta_i}(\xi_i))\cap B_R(0)}|\nabla\psi|^2_{\mathrm{Eucl}}\,dx}\,.\end{aligned} \tag{41}$$

Via Lebesque's theorem, it follows from (40) that for $R > 0$ fixed, the right-hand side above is as small as desired for $N > 0$ large. Plugging together (38), (39), and (41), we get that for $i \neq j$,

$$\left| \int_M (\nabla \operatorname{Resc}^F_{\delta_i,\xi_i}(\varphi), \nabla \operatorname{Resc}^F_{\delta_j,\xi_j}(\psi))_g \, dv_g \right| \leq \omega_{7,F,\varphi,\psi}(N), \tag{42}$$

where $\lim_{N\to+\infty} \omega_{7,F,\varphi,\psi}(N) = 0$.

We now assume that $i = j$. For $R > 0$ fixed, we have that $|g_{\xi_i} - g| \leq C(R)\delta_i g$ on $B_{R\delta_i}(\xi_i)$ since $\Lambda_{\xi_i}(\xi_i) = 1$. We then get with (26) that

$$\begin{aligned} &\left| \int_{B^{g_{\xi_i}}_{R\delta_i}(\xi_i)} (\nabla \operatorname{Resc}^F_{\delta_i,\xi_i}(\varphi), \nabla \operatorname{Resc}^F_{\xi_i,\delta_i}(\psi))_g \, dv_g \right. \\ &\quad \left. - \int_{B^{g_{\xi_i}}_{R\delta_i}(\xi_i)} (\nabla \operatorname{Resc}^F_{\delta_i,\xi_i}(\varphi), \nabla \operatorname{Resc}^F_{\xi_i,\delta_i}(\psi))_{g_{\xi_i}} \, dv_{g_{\xi_i}} \right| \\ &\leq C(F,R)\delta_i \| \operatorname{Resc}^F_{\delta_i,\xi_i}(\varphi)\|_{H_1^2} \| \operatorname{Resc}^F_{\delta_j,\xi_j}(\psi)\|_{H_1^2} \leq C(F,R,\varphi,\psi)\delta_i\,. \end{aligned} \tag{43}$$

Proposition 3.2 then follows from (42), (39), (27), (29), and (43). □

As a corollary, we get an orthogonality property for the $Z_{\delta_i,\xi_i,j}$'s defined in (19):

Corollary 3.3. *Let $h \in L^\infty(M)$ be such that $\|h\|_\infty \leq \tilde{C}_1$. For any $i, i' \in \{1, \ldots, k\}$ and any $j, j' \in \{0, \ldots, n\}$, we have that*

$$\left| (Z_{\delta_i,\xi_i,j}, Z_{\delta_{i'},\xi_{i'},j'})_h - \delta_{i,i'}\delta_{j,j'} \|\nabla V_j\|_2^2 \right| \leq \omega_{8,\tilde{C}_1}(\varepsilon, N),$$

where $\lim_{\varepsilon\to 0,\, N\to+\infty} \omega_{8,\tilde{C}_1}(\varepsilon, N) = 0$. *Here,* $\delta_{i,i'} = 1$ *if* $i = i'$ *and* 0 *otherwise.*

Proof of Corollary 3.3. Taking $F(\xi, x) := \chi(d_{g_\xi}(x,\xi))\Lambda_\xi(x)$, the corollary is a direct consequence of (25), Proposition 3.4 above, and the fact that the V_j's form an orthogonal family of $D_1^2(\mathbb{R}^n)$. □

We now deal with the nonlinear interactions of different rescalings:

Proposition 3.4. *Let $\varphi, \psi \in D_1^2(\mathbb{R}^n)$ be two functions. Then for any $i \neq j \in \{1, \ldots, k\}$ and all $r, s \geq 0$ such that $1 \leq r + s \leq 2^\star$, we have that*

$$\int_M |\operatorname{Resc}^F_{\delta_i,\xi_i}(\varphi)|^r |\operatorname{Resc}^F_{\delta_j,\xi_j}(\varphi)|^s \, dv_g \leq \omega_{9,F,\varphi,\psi}(\varepsilon, N), \tag{44}$$

where $\lim_{\varepsilon\to 0,\, N\to+\infty} \omega_{9,F,\varphi,\psi}(\varepsilon, N) = 0$.

Proof of Proposition 3.4. We let $R > 0$ be a positive number. We have that

$$\begin{aligned} &\int_M |\operatorname{Resc}^F_{\delta_i,\xi_i}(\varphi)|^r |\operatorname{Resc}^F_{\delta_j,\xi_j}(\psi)|^s \, dv_g \\ &\leq \int_{M\setminus B^{g_{\xi_i}}_{R\delta_i}(\xi_i)} \cdots + \int_{M\setminus B^{g_{\xi_j}}_{R\delta_j}(\xi_j)} \cdots + \int_{B^{g_{\xi_i}}_{R\delta_i}(\xi_i)\cap B^{g_{\xi_j}}_{R\delta_j}(\xi_j)} \cdots \end{aligned} \tag{45}$$

It follows from (11), Hölder's inequality, (26), and Sobolev's embedding (22) that

$$\int_{M\setminus B^{g_{\xi_i}}_{R\delta_i}(\xi_i)} |\operatorname{Resc}^F_{\delta_i,\xi_i}(\varphi)|^r |\operatorname{Resc}^F_{\delta_j,\xi_j}(\psi)|^s \, dv_g \le (\operatorname{Vol}_g(M))^{\frac{2^\star-(r+s)}{2^\star}} \tag{46}$$

$$\times \left(\int_{M\setminus B^{g_{\xi_i}}_{R\delta_i}(\xi_i)} |\operatorname{Resc}^F_{\delta_i,\xi_i}(\varphi)|^{2^\star}\, dv_{g_\xi}\right)^{\frac{r}{2^\star}} \cdot \|\operatorname{Resc}^F_{\delta_j,\xi_j}(\psi)\|^s_{2^\star}$$

$$\le (\operatorname{Vol}_g(M))^{\frac{2^\star-(r+s)}{2^\star}} \left(\int_{\mathbb{R}^n\setminus B_R(0)} |\varphi|^{2^\star}\, dx\right)^{\frac{r}{2^\star}} \cdot C(F)\|\operatorname{Resc}^F_{\delta_j,\xi_j}(\psi)\|^s_{H^2_1}$$

$$\le C(F)\cdot(1+\operatorname{Vol}_g(M))\cdot\left(\int_{\mathbb{R}^n\setminus B_R(0)} |\varphi|^{2^\star}\, dx\right)^{\frac{r}{2^\star}} \cdot \|\psi\|^s_{D^2_1} \le \omega_{10,F,\varphi,\psi}(R),$$

where $\lim_{R\to+\infty}\omega_{10,F,\varphi,\psi}(R)=0$.

We now assume that $B^{g_{\xi_i}}_{R\delta_i}(\xi_i)\cap B^{g_{\xi_j}}_{R\delta_j}(\xi_j)\neq\emptyset$ and $i\neq j$. Then we have that $d_g(\xi_i,\xi_j) < C_1R(\delta_i+\delta_j)$, where $C_1>0$. Since $((\delta_i)_i,(\xi_i)_i)\in\mathcal{D}_k(\varepsilon,N)$, up to exchanging i and j if necessary, we then get that

$$\frac{\delta_i}{\delta_j} < \frac{2(1+C_1^2R^2)}{N}. \tag{47}$$

Therefore, using the comparison between g_{ξ_i} and g_{ξ_j} given by (11), we get that

$$\int_{B^{g_{\xi_i}}_{R\delta_i}(\xi_i)\cap B^{g_{\xi_j}}_{R\delta_j}(\xi_j)} |\operatorname{Resc}^F_{\delta_i,\xi_i}(\varphi)|^r |\operatorname{Resc}^F_{\delta_j,\xi_j}(\psi)|^s\, dv_g$$

$$\le (\operatorname{Vol}_g(M))^{\frac{2^\star-(r+s)}{2^\star}} \|\operatorname{Resc}^F_{\delta_i,\xi_i}(\varphi)\|^r_{L^{2^\star}(M)}$$

$$\times\left(\int_{B^{g_{\xi_j}}_{C_8R\delta_i}(\xi_i)\cap B^{g_{\xi_j}}_{R\delta_j}(\xi_j)} |\operatorname{Resc}^F_{\delta_j,\xi_j}(\psi)|^{2^\star}\, dv_{g_\xi}\right)^{\frac{s}{2^\star}}$$

$$\le C(F)\|\operatorname{Resc}^F_{\delta_i,\xi_i}(\varphi)\|_{H^2_1}\cdot\left(\int_{\delta_j^{-1}\tilde{\exp}^{-1}_{\xi_j}(B^{g_{\xi_j}}_{C_2R\delta_i}(\xi_i))\cap B_R(0)} |\psi|^{2^\star}\, dv_{g_\xi}\right)^{\frac{s}{2^\star}}$$

$$\le C(F)\|\varphi\|_{D^2_1}\cdot\left(\int_{\delta_j^{-1}\tilde{\exp}^{-1}_{\xi_j}(B^{g_{\xi_j}}_{C_2R\delta_i}(\xi_i))\cap B_R(0)} |\psi|^{2^\star}\, dv_{g_\xi}\right)^{\frac{s}{2^\star}}. \tag{48}$$

Via Lebesque's theorem, it follows from (47) that for $R>0$ fixed, the right-hand side above is as small as desired for $N>0$ large. Plugging (46) and (48) into (45) yields (44). This ends the proof of Proposition 3.4. □

The last tool introduced here is the inverse rescaling. Let $\tilde{F}\in C^\infty(M\times\mathbb{R}^n)$ be such that $\tilde{F}(\xi,z)=0$ if $|z|\ge r_0$. Let $\phi\in H^2_1(M)$ be a function. For $\xi\in U_\gamma$ and

$\delta > 0$, we define

$$\tilde{\operatorname{Resc}}^{\tilde F}_{\delta,\xi}(\phi)(x) := \tilde F(\xi, \delta|x|)\delta^{\frac{n-2}{2}}\phi \circ \tilde{\exp}_\xi(\delta x) \tag{49}$$

for all $x \in \mathbb{R}^n$.

Proposition 3.5. *For any $\phi \in H_1^2(M)$, $\xi \in U_\gamma$, and $\delta > 0$, then $\tilde{\operatorname{Resc}}^{\tilde F}_{\delta,\xi}(\phi) \in D_1^2(\mathbb{R}^n)$. In addition, if $\|h\|_\infty \leq \tilde C_1$, then*

$$\|\tilde{\operatorname{Resc}}^{\tilde F}_{\delta,\xi}(\phi)\|_{D_1^2} \leq C(\tilde C_1)\|\phi\|_{H_1^2(M)}.$$

Proof of Proposition 3.5. By density, it is enough to prove the result for $\phi \in C^\infty(M)$. Then, $\tilde{\operatorname{Resc}}^{\tilde F}_{\delta,\xi}(\phi) \in C_c^\infty(\mathbb{R}^n)$. A change of variable yields

$$\begin{aligned}&\int_{\mathbb{R}^n} |\nabla \tilde{\operatorname{Resc}}^{\tilde F}_{\delta,\xi}(\phi)|^2_{\mathrm{Eucl}}\, dx \\ &= \int_{B^{g_\xi}_{r_0}(\xi)} |\nabla(\tilde F(\xi, (\tilde{\exp}^{g_\xi}_\xi)^{-1}))\phi|^2_{(\tilde{\exp}^{g_\xi}_\xi)^{-1})^\star \mathrm{Eucl}}\, dv_{(\tilde{\exp}^{g_\xi}_\xi)^{-1})^\star \mathrm{Eucl}}.\end{aligned}$$

Since $F \in C^\infty(M \times \mathbb{R}^n)$ and (11) holds, we have that $(\tilde{\exp}^{g_\xi}_\xi)^{-1})^\star\,\mathrm{Eucl} \leq Cg$ and

$$\int_{\mathbb{R}^n} |\nabla \tilde{\operatorname{Resc}}^{\tilde F}_{\delta,\xi}(\phi)|^2_{\mathrm{Eucl}}\, dx \leq C(\tilde F)\|\phi\|^2_{H_1^2(M)}$$

for all $\phi \in C^\infty(M)$. Proposition 3.5 then follows by density. □

4. Preliminary computations 2: Estimates of derivatives

This section is devoted to the proof of the following estimates:

Proposition 4.1. *For $\gamma \in \mathcal{C}$, for any $\xi \in U_\gamma$ and $\delta > 0$, we have that*

$$\partial_\delta W_{\kappa,\delta,\xi} = \kappa \frac{n-2}{2}\left(\frac{n(n-2)}{f_0(\xi)}\right)^{\frac{n-2}{4}} \cdot \frac{1}{\delta}\cdot (Z_{\xi,\delta,0} + o(1)), \tag{50}$$

$$\partial_{(\xi)_j} W_{\kappa,\delta,\xi} = \kappa \frac{n-2}{2}\left(\frac{n(n-2)}{f_0(\xi)}\right)^{\frac{n-2}{4}} \cdot \frac{1}{\delta}\cdot (Z_{\xi,\delta,j} + o(1)) \tag{51}$$

for all $j = 1, \dots, n$, where $\|o(1)\|_{H_1^2} \leq \omega_{11}(\delta)$ and $\lim_{\delta\to 0}\omega_{11}(\delta) = 0$. Moreover, we have that

$$\delta\|\partial_\delta Z_{\xi,\delta,j}\|_{H_1^2} \leq C \text{ and } \delta\|\nabla_\xi Z_{\xi,\delta,j}\|_{H_1^2} \leq C, \tag{52}$$

where $C > 0$ is independent of $\xi \in U_\gamma$ and $\delta \in (0,1)$. The partial derivatives along the center $\xi \in U_\gamma$ in (51) are defined in (56) below.

In other words, the differentiation of the rescaling along $(0,+\infty)\times M$ is essentially the rescaling of the differentiation of U_1 along the Lie algebra of $(0,+\infty)\times\mathbb{R}^n$ for the action (15).

Proof of Proposition 4.1. Straightforward computations yield

$$\partial_\delta W_{\kappa,\delta,\xi} = \kappa\frac{n-2}{2}\left(\frac{n(n-2)}{f_0(\xi)}\right)^{\frac{n-2}{4}}\cdot\frac{1}{\delta}\cdot Z_{\xi,\delta,0}+\partial_\delta B_{\delta,\xi}\,, \tag{53}$$

$$\partial_\delta Z_{\xi,\delta,j} = \frac{1}{\delta}\cdot \mathrm{Resc}_{\delta,\xi}^{F^{(1)}}(\Phi_j) \text{ for all } j=0,\dots,n, \tag{54}$$

where $F^{(1)}(\xi,x) := \chi(d_{g_\xi}(x,\xi))\Lambda_\xi(x)$ for $\xi, x\in M$ and $\Phi_j\in D_1^2(\mathbb{R}^n)$ are such that

$$\Phi_0(x) := \frac{\frac{n-2}{2}|x|^4-(n+2)|x|^2+\frac{n-2}{2}}{(1+|x|^2)^{\frac{n+2}{2}}}\,,$$

$$\Phi_j(x) := \frac{n(|x|^2-1)x_j}{2(1+|x|^2)^{\frac{n+2}{2}}} \text{ for } j=1,\dots,n$$

for all $x\in\mathbb{R}^n$. It then follows from (13), (53), (26), (54) that (50) and the first inequality of (52) hold.

We now focus on the derivatives along the center ξ. Since the $W_{\delta,\xi}$'s and the $Z_{\delta,\xi,j}$'s enjoy the same representation (25), we work with the function

$$\mathcal{W}_{\delta,\xi}(x) := \Psi(\xi,x)\delta^{-\frac{n-2}{2}}V\big(\delta^{-1}(\tilde{\exp}_\xi^{g_\xi})^{-1}(x)\big)$$

for all $x\in M$, where $V\in D_1^2(\mathbb{R}^n)$ is such that $\partial_j V\in D_1^2(\mathbb{R}^n)$ for all $j=1,\dots,n$ and $\Psi\in C^\infty(U_\gamma\times M)$ is such that $\Psi(\xi,x)=0$ if $d_{g_\xi}(x,\xi)\ge r_0$. For $\vec{\tau}\in\mathbb{R}^n$, we define $\vec{\tau}(t) := \tilde{\exp}_\xi^{g_\xi}(t\vec{\tau})$, and we consider

$$\mathcal{W}_{\delta,\vec{\tau}(t)} := \Psi(\vec{\tau}(t),\cdot)\delta^{-\frac{n-2}{2}}V\left(\delta^{-1}\Theta_\xi\right), \tag{55}$$

where $\Theta_\xi := (\tilde{\exp}_\xi^{g_\xi})^{-1}$ with definition (20). We then define

$$\partial_{(\xi)_j}\mathcal{W}_{\delta,\xi} := \frac{d}{dt}\left(\mathcal{W}_{\delta,\vec{\tau}(t)}\right)_{|t=0} \tag{56}$$

for $\vec{\tau}=(0,\dots,0,1,0,\dots,0)$ being the jth vector of the canonical basis of $\mathbb{R}^n$. Straightforward computations yield

$$\frac{d}{dt}\left(\mathcal{W}_{\delta,\vec{\tau}(t)}\right)_{|t=0} = \mathrm{Resc}_{\delta,\xi}^{\Psi_0}(V)+\sum_{k=1}^n \delta^{-1}\,\mathrm{Resc}_{\delta,\xi}^{\Psi_k}(\partial_k V),$$

where

$$\Psi_0(\xi,x) := d\Psi_{(\xi,x)}(d(\tilde{\exp}_\xi^{g_\xi})_0(\vec{\tau}),0),$$

$$\Psi_k(\xi,x) := \Psi(\xi,x)\frac{d}{dt}\left(\Theta_{\vec{\tau}(t)}(x)\right)^k_{|t=0}$$

for $k=1,\dots,n$, $\xi\in U_\gamma$, and $x\in M$. We define

$$\tilde{\Psi}_k(\xi,x) := \Psi_k(\xi,x)-\Psi_k(\xi,\xi)F^{(1)}(\xi,x),$$

where

$$F^{(1)}(\xi,x) := \chi(d_{g_\xi}(x,\xi))\Lambda_\xi(x)$$

for all $k = 1, \dots, n$ and $\xi, x \in M$. We then have that

$$\frac{d}{dt}\left(\mathcal{W}_{\delta,\vec{\tau}(t)}\right)_{|t=0} = \sum_{k=1}^{n} \delta^{-1}\Psi_k(\xi,\xi)\,\mathrm{Resc}_{\delta,\xi}^{F^{(1)}}(\partial_k V) + \mathrm{Resc}_{\delta,\xi}^{\Psi_0}(V) + \sum_{k=1}^{n} \delta^{-1}\,\mathrm{Resc}_{\delta,\xi}^{\tilde{\Psi}_k}(\partial_k V).$$

Since $\tilde{\Psi}_k(\xi,\xi) = 0$, it follows from Proposition 3.1 that

$$\left\| \frac{d}{dt}\left(\mathcal{W}_{\delta,\vec{\tau}(t)}\right)_{|t=0} - \sum_{k=1}^{n} \delta^{-1}\Psi_k(\xi,\xi)\,\mathrm{Resc}_{\delta,\xi}^{F^{(1)}}(\partial_k V)\right\|_{H_1^2} \leq \delta^{-1}\omega_{12,V}(\delta), \tag{57}$$

where $\lim_{\delta\to 0}\omega_{12,V}(\delta) = 0$. We are left with computing $\Psi_k(\xi,\xi)$. We define $X(t) := \Theta_{\vec{\tau}(t)}(\xi)$ and for t small. In particular, X and ξ are smooth with respect to t small. The definition of Θ, Taylor expansions, and the fact that $X(0) = 0$ yield

$$\begin{aligned}
0 &= \left(\tilde{\exp}_\xi^{g_\xi}\right)^{-1} \circ \tilde{\exp}_{\vec{\tau}(t)}^{g_{\vec{\tau}(t)}}(X(t)) = \left(\tilde{\exp}_\xi^{g_\xi}\right)^{-1} \circ \tilde{\exp}_{\vec{\tau}(t)}^{g_{\vec{\tau}(t)}}(tX'(0) + o(t))\\
&= \left(\tilde{\exp}_\xi^{g_\xi}\right)^{-1} \circ \tilde{\exp}_{\vec{\tau}(t)}^{g_{\vec{\tau}(t)}}(0) + td\left(\left(\tilde{\exp}_\xi^{g_\xi}\right)^{-1} \circ \tilde{\exp}_{\vec{\tau}(t)}^{g_{\vec{\tau}(t)}}\right)_0(X'(0)) + o(t)\\
&= \left(\tilde{\exp}_\xi^{g_\xi}\right)^{-1}(\vec{\tau}(t)) + td\left(\left(\tilde{\exp}_\xi^{g_\xi}\right)^{-1} \circ \tilde{\exp}_\xi^{g_\xi}\right)_0(X'(0)) + o(t)\\
&= t\vec{\tau} + tX'(0) + o(t)
\end{aligned}$$

when $t \to 0$. Therefore $X'(0) = -\vec{\tau}$. Since $\vec{\tau} = (0, \dots, 0, 1, 0, \dots,)$ (the jth vector), we then get that $\Psi_k(\xi,\xi) = -\Psi(\xi,\xi)$ if $k = j$ and 0 otherwise. Then (57) rewrites

$$\left\| \frac{d}{dt}\left(\mathcal{W}_{\delta,\tilde{\exp}_\xi^{g_\xi}(t\vec{\tau})}\right)_{|t=0} - \delta^{-1}\Psi(\xi,\xi)\,\mathrm{Resc}_{\delta,\xi}^{F^{(1)}}(\partial_j V)\right\|_{H_1^2} \leq \delta^{-1}\omega_{12,V}(\delta), \tag{58}$$

where $\lim_{\delta\to 0}\omega_{12}(\delta) = 0$. The assertion (51) and the second assertion of (52) follow from the expressions (25) and (55), and from (13) and (58). □

5. Inversion and fixed-point argument

For $((\delta_i)_i, (\xi_i)_i) \in (0, +\infty)^k \times M^k$, we define

$$K_{(\delta_i)_i,(\xi_i)_i} := \mathrm{Span}\left\{Z_{\delta_i,\xi_i}\,;\, Z_{\delta_i,\xi_i,\omega_i}\,;\, \varphi \,/\, i = 1, \dots, k\,,\, \omega_i \in T_{\xi_i}M\,,\, \varphi \in K_0\right\}.$$

We let $\{\varphi_1, \dots, \varphi_d\}$ be an orthonormal basis of K_0 for $(\cdot,\cdot)_{h_0}$. We then have that

$$K_{(\delta_i)_i,(\xi_i)_i} = \mathrm{Span}\left\{Z_{\delta_i,\xi_i,j}\,;\, \varphi_l \,/\, i = 1, \dots, k\,,\, j = 0, \dots, n\,,\, l = 1, \dots, d\right\}. \tag{59}$$

It follows from (25), (26), and (29) that the $Z_{\delta_i,\xi_i,j}$'s go weakly to 0 in $H_1^2(M)$ when $\delta_i \to 0$ uniformly with respect to $\xi_i \in U_{\gamma(i)}$. It follows from Corollary 3.3 that for $\varepsilon > 0$ small enough and $N > 0$ large enough, the $Z_{\delta_i,\xi_i,j}$'s ($i = 1, \dots, k$ and $j = 0, \dots, n$) form an "almost" orthogonal family. Therefore, the generating family in (59) is "almost" orthogonal for $(\delta_i)_i, (\xi_i)_i \in \mathcal{D}_k(\varepsilon, N)$ for $\varepsilon > 0$ small and $N > 0$ large, and therefore, $\dim_{\mathbb{R}} K_{(\delta_i)_i,(\xi_i)_i} = k(n+1) + d$. We define $K_{(\delta_i)_i,(\xi_i)_i}^{\perp}$ as the orthogonal of $K_{(\delta_i)_i,(\xi_i)_i}$ in $H_1^2(M)$ for the scalar product $(\cdot,\cdot)_h$.

We define $\Pi_{K(\delta_i)_i,(\xi_i)_i} : H_1^2(M) \to H_1^2(M)$ and $\Pi_{K^\perp_{(\delta_i)_i,(\xi_i)_i}} : H_1^2(M) \to H_1^2(M)$ respectively as the orthogonal projection on $K_{(\delta_i)_i,(\xi_i)_i}$ and $K^\perp_{(\delta_i)_i,(\xi_i)_i}$ with respect to the scalar product $(\cdot,\cdot)_h$. As easily checked, $v \in H_1^2(M)$ is a solution to (1) if and only if

$$\begin{aligned} &\Pi_{K^\perp_{(\delta_i)_i,(\xi_i)_i}}\left(v-(\Delta_g+h)^{-1}(F'(v))\right)=0\\ \text{and }\ &\Pi_{K_{(\delta_i)_i,(\xi_i)_i}}\left(v-(\Delta_g+h)^{-1}(F'(v))\right)=0\,. \end{aligned} \tag{60}$$

In this section, we solve the first equation of (60):

Proposition 5.1. *Under the hypotheses of Theorem 1.1, there exists $N>0$ and $\varepsilon>0$ such that for any $h \in L^\infty(M)$, $f \in C^0(M)$, $G \in C^{2,\theta}_{\mathrm{loc}}(H_1^2(M))$, and $u \in C^1(B_1(0),H_1^2(M))$ such that (6), (7), and (8) hold, there exists $\phi \in C^1(B_\varepsilon(0)\times \mathcal{D}_k(\varepsilon,N),H_1^2(M))$ such that*

$$u(z,(\delta_i)_i,(\xi_i)_i) := u(z) + \sum_{i=1}^k W_{\kappa_i,\delta_i,\xi_i} + \phi(z,(\delta_i)_i,(\xi_i)_i)$$

is a solution to

$$\Pi_{K^\perp_{(\delta_i)_i,(\xi_i)_i}}\left(u(z,(\delta_i)_i,(\xi_i)_i)-(\Delta_g+h)^{-1}(F'(u(z,(\delta_i)_i,(\xi_i)_i)))\right)=0 \tag{61}$$

for all $(z,(\delta_i)_i,(\xi_i)_i) \in B_\varepsilon(0)\times\mathcal{D}_k(\varepsilon,N)$. In addition, we have that

$$\phi(z,(\delta_i)_i,(\xi_i)_i) \in K^\perp_{(\delta_i)_i,(\xi_i)_i} \text{ and } \|\phi(z,(\delta_i)_i,(\xi_i)_i)\|_{H_1^2} \le C\cdot R(z,(\delta_i)_i,(\xi_i)_i),$$

where C is a constant depending on (M,g), k, ν_0, θ, C_0, u_0, h_0, f_0, and G_0. The remainder $R(z,(\delta_i)_i,(\xi_i)_i)$ is defined in (5). Moreover, we have that

$$R(z,(\delta_i)_i,(\xi_i)_i) \le \omega_{13}(\varepsilon,N)$$

for all $z\in B_\varepsilon(0)$, $((\delta_i)_i,(\xi_i)_i)\in\mathcal{D}_k(\varepsilon,N)$, where $\lim_{\varepsilon\to0,N\to+\infty}\omega_{13}(\varepsilon,N)=0$.

5.1. Inversion of the linearized operator

Proposition 5.2. *Under the hypotheses of Theorem 1.1, there exists $c>0$, there exists $N>0$ and $\varepsilon>0$ such that for $h\in L^\infty(M)$, $f\in C^0(M)$, $G\in C^{2,\theta}_{\mathrm{loc}}(H_1^2(M))$, and $u\in C^1(B_1(0),H_1^2(M))$ such that (6), (7), and (8) hold, then there exists $c>0$ such that for any $z\in B_\varepsilon(0)$ and $((\delta_i)_i,(\xi_i)_i)\in\mathcal{D}_k(\varepsilon,N)$, we have that*

$$\|L_{z,(\delta_i)_i,(\xi_i)_i}(\varphi)\|_{H_1^2} \ge c\|\varphi\|_{H_1^2} \tag{62}$$

for all $\varphi\in H_1^2(M)$, where

$$L_{z,(\delta_i)_i,(\xi_i)_i)}:\left\{\begin{array}{ccc} K^\perp_{(\delta_i)_i,(\xi_i)_i} & \to & K^\perp_{(\delta_i)_i,(\xi_i)_i} \\ \varphi & \mapsto & \Pi_{K^\perp_{(\delta_i)_i,(\xi_i)_i}}\left(\varphi-(\Delta_g+h)^{-1}(F''(u(z,(\delta_i)_i,(\xi_i)_i)\varphi)\right) \end{array}\right\}.$$

In particular, $L_{z,(\delta_i)_i,(\xi_i)_i)}$ is a bi-continuous isomorphism.

Proof of Proposition 5.2. We prove (62) by contradiction. We assume that there exist $(q_\alpha)_\alpha \in (2, 2^\star]$, $(h_\alpha)_\alpha \in L^\infty(M)$, $(f_\alpha)_\alpha \in C^0(M)$, $(z_\alpha)_\alpha \in B_1(0)$, $(u_\alpha)_\alpha \in C^1(B_1(0); H_1^2(M))$, $(G_\alpha)_\alpha \in C^{2,\theta}_{\text{loc}}(H_1^2(M))$, $(\delta_{i,\alpha})_\alpha$, and $(\xi_{i,\alpha})_\alpha$ for $i = 1, \dots, k$ and $(\phi_\alpha)_\alpha \in H_1^2(M)$ such that

$$\lim_{\alpha\to+\infty} \|h_\alpha - h_0\|_\infty + \|f_\alpha - f_0\|_{C^0} + d_{C_B^{2,\theta}}(G_\alpha, G_0) = 0\,, \tag{63}$$

$$\lim_{\alpha\to+\infty} \delta_{i,\alpha} = 0\,,\ \lim_{\alpha\to+\infty} \delta_{i,\alpha}^{2^\star - q_\alpha} = 1\,,\ \lim_{\alpha\to+\infty} z_\alpha = 0\,,\ \lim_{\alpha\to+\infty} q_\alpha = 2^\star\,, \tag{64}$$

$$u_\alpha(0) = u_0\,,\ \|u_\alpha\|_{C^1(B_1(0), H_1^2)} \le C_0\,,\ f_0(\xi_{i,\alpha}) \ge \nu_0 \text{ for all } i = 1, \dots, k, \tag{65}$$

$$\lim_{\alpha\to+\infty} \left(\frac{\delta_{i,\alpha}}{\delta_{j,\alpha}} + \frac{\delta_{j,\alpha}}{\delta_{i,\alpha}} + \frac{d_g(\xi_{i,\alpha}, \xi_{j,\alpha})^2}{\delta_{i,\alpha}\delta_{j,\alpha}} \right) = +\infty\,, \tag{66}$$

$$\|\phi_\alpha\|_{H_1^2} = 1\,,\ \phi_\alpha \in K_\alpha^\perp\,, \tag{67}$$

and

$$L_\alpha(\phi_\alpha) = o(1), \tag{68}$$

where $\lim_{\alpha\to+\infty} o(1) = 0$ in $H_1^2(M)$ and

$$L_\alpha := L_{z_\alpha, (\delta_{i,\alpha})_i, (\xi_{i,\alpha})_i} \text{ and } K_\alpha := K_{(\delta_{i,\alpha})_i, (\xi_{i,\alpha})_i}\,.$$

In the sequel, all convergences are with respect to a subsequence of α. It follows from the boundedness of $(\phi_\alpha)_\alpha$ in (67) that there exists $\phi \in H_1^2(M)$ such that

$$\phi_\alpha \rightharpoonup \phi \text{ weakly in } H_1^2(M) \text{ when } \alpha \to +\infty\,. \tag{69}$$

It follows from (68) that there exist $(\lambda_\alpha^{ij})_\alpha \in \mathbb{R}$ and $(\mu_\alpha^l)_\alpha \in \mathbb{R}$ for $i \in \{1, \dots, k\}$, $j \in \{0, \dots, n\}$, and $l \in \{1, \dots, d\}$ such that

$$\phi_\alpha - (\Delta_g + h_\alpha)^{-1}\Big(F_\alpha''\Big(u_\alpha(z_\alpha) + \sum_{i=1}^k W_{i,\alpha}\Big)\phi_\alpha\Big) = o(1) + \sum_{i,j} \lambda_\alpha^{ij} Z_{i,j,\alpha} + \sum_{l=1}^d \mu_\alpha^l \varphi_l\,, \tag{70}$$

where $\lim_{\alpha\to 0} o(1) = 0$ in $H_1^2(M)$ and

$$F_\alpha(v) := \frac{1}{q_\alpha}\int_M f_\alpha H(v)^{q_\alpha}\, dv_g + G_\alpha(v) \tag{71}$$

for all $v \in H_1^2(M)$ and

$$W_{i,\alpha} := W_{\kappa_i, \delta_{i,\alpha}, \xi_{i,\alpha}} \text{ and } Z_{i,j,\alpha} := Z_{\delta_{i,\alpha}, \xi_{i,\alpha}, j}$$

for all $i \in \{1, \dots, k\}$ and $j \in \{0, \dots, n\}$. It follows from Proposition 3.1 that for any $i \in \{1, \dots, k\}$ and $j \in \{0, \dots, n\}$, we have that

$$W_{i,\alpha} \rightharpoonup 0 \text{ and } Z_{i,j,\alpha} \rightharpoonup 0 \text{ weakly in } H_1^2(M) \text{ when } \alpha \to +\infty\,. \tag{72}$$

Since $\phi_\alpha \in K_\alpha^\perp$, for any $i = 1, \dots, k$ and any $j = 0, \dots, n$, we have that

$$(\phi_\alpha, Z_{i,j,\alpha})_{h_\alpha} = 0 \text{ and } (\phi_\alpha, \varphi)_{h_\alpha} = 0 \text{ for all } \varphi \in K_0\,. \tag{73}$$

It follows from the local C^2-convergence (63) of G_α to G_0, from the continuity properties of G_0 (see Definition 2.1) and from (69) that

$$\begin{aligned}\phi_\alpha - (\Delta_g + h_\alpha)^{-1}\Big((q_\alpha - 1) f_\alpha H\Big(u_\alpha(z_\alpha) + \sum_{i=1}^{k} W_{i,\alpha}\Big)^{q_\alpha - 2}\phi_\alpha\Big) \\ - (\Delta_g + h_0)^{-1}(G_0''(u_0)\phi) = o(1) + \sum_{i,j}\lambda_\alpha^{ij} Z_{i,j,\alpha} + \sum_{l=1}^{d}\mu_\alpha^l \varphi_l\end{aligned} \tag{74}$$

where $\lim_{\alpha\to 0} o(1) = 0$ in $H_1^2(M)$. We define

$$\Lambda_\alpha := \sum_{i=1}^{k}\sum_{j=0}^{n}|\lambda_\alpha^{ij}| + \sum_{l=1}^{d}|\mu_\alpha^l| \text{ for all } \alpha.$$

We fix $\varphi \in H_1^2(M)$. It then follows from (74) that

$$(\phi_\alpha, \varphi)_{h_\alpha} - (q_\alpha - 1)\int_M f_\alpha H\Big(u_\alpha(z_\alpha) + \sum_{i=1}^{k} W_{i,\alpha}\Big)^{q_\alpha - 2}\phi_\alpha\varphi\, dv_g - G_0''(u_0)(\phi,\varphi)$$

$$= o_\alpha(1)\,(1+\Lambda_\alpha))\,(\|\varphi\|_{H_1^2}) + \sum_{i,j}\lambda_\alpha^{ij}(Z_{i,j,\alpha},\varphi)_{h_\alpha} + \sum_{l=1}^{d}\mu_\alpha^l(\varphi_l,\varphi)_{h_0}\,. \tag{75}$$

Here and in the sequel, $\lim_{\alpha\to+\infty} o_\alpha(1) \to 0$ uniformly with respect to $\varphi \in H_1^2(M)$.

Step 1: We first bound the μ_α^l's. We fix $\varphi \in H_1^2(M)$. It follows from (63), (64) (65), (25), and (26) that the family $(f_\alpha H(u_\alpha(z_\alpha) + \sum_{i=1}^k W_{i,\alpha})^{q_\alpha-2}\phi_\alpha)_\alpha$ is uniformly bounded in $L^{2n/(n+2)}(M)$ and converges a.e. to $f_0 H(u_0)^{2^\star-2}\phi$ when $\alpha \to +\infty$. It then follows from integration theory that the convergence holds weakly in $L^{2^\star}(M)'$. Therefore, passing to the limit $\alpha \to +\infty$ in (75) for $\varphi \in H_1^2(M)$ fixed, we get that

$$\begin{aligned}(\phi,\varphi)_{h_0} - (2^\star - 1)\int_M f_0 H(u_0)^{2^\star-2}\phi\varphi\, dv_g - G_0''(u_0)(\phi,\varphi) \\ = (\phi,\varphi)_{h_0} - F_0''(u_0)(\phi,\varphi) = o_\alpha(1)\,(1+\Lambda_\alpha) + \sum_{l=1}^{d}\mu_\alpha^l(\varphi_l,\varphi)_{h_0}\,.\end{aligned} \tag{76}$$

Passing to the limit $\alpha \to +\infty$ in the second equality of (73) yields $(\phi,\varphi)_{h_0} = 0$ for all $\varphi \in K_0$. It then follows from (3) that $F_0''(u_0)(\varphi,\phi) = (\phi,\varphi)_{h_0} = 0$, and then

$$\sum_{l=1}^{d}\mu_\alpha^l(\varphi_l,\varphi)_{h_0} = o_\alpha(1)\,(1+\Lambda_\alpha)$$

for all $\varphi \in K_0$. Since $\{\varphi_l /\, l = 1,\ldots,d\}$ is an orthonormal basis of K_0, we get that

$$\sum_{l=1}^{d}|\mu_\alpha^l| = o_\alpha(1)\,(1+\Lambda_\alpha)\,, \tag{77}$$

where $\lim_{\alpha\to+\infty} o_\alpha(1) = 0$.

Step 2: We bound the λ_α^{ij}'s. We fix $i_0 \in \{1, \dots, k\}$. We fix $\varphi \in D_1^2(\mathbb{R}^n)$ and define

$$\varphi_{i_0,\alpha} := \mathrm{Resc}^F_{\xi_{i_0,\alpha},\delta_{i_0,\alpha}}(\varphi), \text{ where } F(\xi, x) := \chi(d_{g_\xi}(\xi, x))\Lambda_\xi(x) \text{ for } \xi, x \in M.$$

In particular, it follows from Proposition 3.1 that

$$\varphi_{i_0,\alpha} \rightharpoonup 0 \text{ weakly in } H_1^2(M) \text{ when } \alpha \to +\infty. \tag{78}$$

We define

$$\tilde{\phi}_{i_0,\alpha} := \tilde{\mathrm{Resc}}^{\tilde{F}}_{\xi_{i_0,\alpha},\delta_{i_0,\alpha}}(\phi_\alpha),$$

where

$$\tilde{F}(\xi, x) := \frac{\chi(|x|)}{\Lambda_\xi(\tilde{\exp}_\xi^{g_\xi}(x))} \text{ for } \xi \in M,\ x \in \mathbb{R}^n.$$

Note that it follows from Proposition 3.5 that $(\tilde{\phi}_{i_0,\alpha})_\alpha$ is bounded in $D_1^2(\mathbb{R}^n)$, and then, there exists $\tilde{\phi}_{i_0} \in D_1^2(\mathbb{R}^n)$ such that

$$\tilde{\phi}_{i_0,\alpha} \rightharpoonup \tilde{\phi}_{i_0} \text{ in } D_1^2(\mathbb{R}^n) \text{ when } \alpha \to +\infty. \tag{79}$$

As easily checked, $\mathrm{Resc}^F_{\xi_{i_0,\alpha},\delta_{i_0,\alpha}}(\tilde{\phi}_{i_0,\alpha}) = \phi_\alpha + \tau_\alpha$, where $(\tau_\alpha)_\alpha$ is bounded in $H_1^2(M)$ with support in $M \setminus B^{g_{\xi_{i_0,\alpha}}}_{\xi_{i_0,\alpha}}(r_0/3)$. It then follows from (28) and (29) that

$$\begin{aligned}\big(\mathrm{Resc}^F_{\xi_{i_0,\alpha},\delta_{i_0,\alpha}}(\tilde{\phi}_{i_0,\alpha}), \mathrm{Resc}^F_{\xi_{i_0,\alpha},\delta_{i_0,\alpha}}(\varphi)\big)_{h_\alpha} \\ = \big(\phi_\alpha, \mathrm{Resc}^F_{\xi_{i_0,\alpha},\delta_{i_0,\alpha}}(\varphi)\big)_{h_\alpha} + o_\alpha(1),\end{aligned} \tag{80}$$

where $\lim_{\alpha\to+\infty} o_\alpha(1) = 0$. Applying (75) to $\varphi_{i_0,\alpha}$ yields

$$\begin{aligned}&(\phi_\alpha, \varphi_{i_0,\alpha})_{h_\alpha} - (q_\alpha - 1)\int_M f_\alpha H\Big(u_\alpha(z_\alpha) + \sum_{i=1}^k W_{i,\alpha}\Big)^{q_\alpha-2}\phi_\alpha\varphi_{i_0,\alpha}\,dv_g \\ &\quad - G_0''(u_0)(\phi, \varphi_{i_0,\alpha}) \\ &= o_\alpha(1)\,(1+\Lambda_\alpha)\,(\|\varphi_{i_0,\alpha}\|_{H_1^2}) + \sum_{i,j}\lambda_\alpha^{ij}(Z_{i,j,\alpha}, \varphi_{i_0,\alpha})_{h_\alpha} + \sum_{l=1}^d \mu_\alpha^l(\varphi_l, \varphi_{i_0,\alpha}))_{h_0}\,.\end{aligned}$$

It then follows from (80), (33), (43), the properties of G_0 (see Definition 2.1), (78), and Proposition 3.2 that

$$\begin{aligned}(\tilde{\phi}_{i_0,\alpha}, \varphi)_{\mathrm{Eucl}} - (q_\alpha - 1)\int_M f_\alpha H\Big(u_\alpha(z_\alpha) + \sum_{i=1}^k W_{i,\alpha}\Big)^{q_\alpha-2}\phi_\alpha\varphi_{i_0,\alpha}\,dv_g \\ = o_\alpha(1)\,(1+\Lambda_\alpha) + \sum_j \lambda_\alpha^{i_0j}(V_j, \varphi)_{\mathrm{Eucl}}\,.\end{aligned} \tag{81}$$

Without loss of generality, we can assume that $\theta < \min\{1, 2^\star - 2\}$. Then, there exists $C(\theta) > 0$ such that

$$\left|H\Big(\sum_{i=0}^k X_i\Big)^{q_\alpha-2} - H(X_0)^{q_\alpha-2}\right| \le C(\theta)|X_0|^\theta \sum_{i\neq 0}^k |X_i|^{q_\alpha-2-\theta} + C(\theta)\sum_{i\neq 0}|X_i|^{q_\alpha-2}$$

for all $X_i \in \mathbb{R}$, $i = 0, \dots, k$. As a consequence, we get that

$$\begin{aligned}
&\left| \int_M f_\alpha \left(H\left(u_\alpha(z_\alpha) + \sum_{i=1}^k W_{i,\alpha} \right)^{q_\alpha - 2} - H(W_{i_0,\alpha})^{q_\alpha-2} \right) \phi_\alpha \varphi_{i_0,\alpha}\, dv_g \right| \\
&\quad \le C \int_M \left(|W_{i_0,\alpha}|^\theta |u_\alpha(z_\alpha)|^{q_\alpha-2-\theta} + \sum_{i \ne i_0} |W_{i_0,\alpha}|^\theta |W_{i,\alpha}|^{q_\alpha-2-\theta} \right) \\
&\qquad \cdot |\phi_\alpha| \cdot |\varphi_{i_0,\alpha}|\, dv_g + \int_M \left(|u_\alpha(z_\alpha)|^{q_\alpha-2} + \sum_{i \ne i_0} |W_{i,\alpha}|^{q_\alpha-2} \right) |\phi_\alpha| \cdot |\varphi_{i_0,\alpha}|\, dv_g \\
&\quad \le C \int_M |W_{i_0,\alpha}|^\theta |\varphi_{i_0,\alpha}| \cdot |\phi_\alpha| \cdot |u_\alpha(z_\alpha)|^{q_\alpha-2-\theta}\, dv_g \\
&\qquad + C \int_M |\varphi_{i_0,\alpha}| \cdot |\phi_\alpha| \cdot |u_\alpha(z_\alpha)|^{q_\alpha-2}\, dv_g \\
&\qquad + C \sum_{i \ne i_0} \| |W_{i_0,\alpha}|^\theta |W_{i,\alpha}|^{(q_\alpha-2-\theta)} \|_{2^\star/(2^\star-2)} \|\varphi_{i_0,\alpha}\|_{2^\star} \|\phi_\alpha\|_{2^\star} \\
&\qquad + C \sum_{i \ne i_0} \| |W_{i,\alpha}|^{q_\alpha-2} \varphi_{i_0,\alpha} \|_{2^\star/(2^\star-1)} \|\phi_\alpha\|_{2^\star}.
\end{aligned}$$

Since $(|W_{i_0,\alpha}|^\theta |\varphi_{i_0,\alpha}| \cdot |\phi_\alpha|)_\alpha$ goes to 0 almost everywhere and is bounded in $L^{2^\star/(2+\theta)}(M)$, since $(|\varphi_{i_0,\alpha}| \cdot |\phi_\alpha|)_\alpha$ goes to 0 almost everywhere and is bounded in $L^{2^\star/2}(M)$, it follows from standard integration theory and Proposition 3.4 that

$$\int_M f_\alpha \left(H\left(u(z_\alpha) + \sum_{i=1}^k W_{i,\alpha} \right)^{q_\alpha-2} - H(W_{i_0,\alpha})^{q_\alpha-2} \right) \phi_\alpha \varphi_{i_0,\alpha}\, dv_g \to 0$$

when $\alpha \to +\infty$. Plugging this limit in (81) yields

$$\begin{aligned}
(\tilde{\phi}_{i_0,\alpha}, \varphi)_{\mathrm{Eucl}} - (q_\alpha - 1) \int_M f_\alpha H\left(W_{i_0,\alpha}\right)^{q_\alpha-2} \phi_\alpha \varphi_{i_0,\alpha}\, dv_g \\
= o_\alpha(1)\left(1 + \Lambda_\alpha\right) + \sum_j \lambda_\alpha^{i_0 j} (V_j, \varphi)_{\mathrm{Eucl}}. \qquad (82)
\end{aligned}$$

For any $R > 0$, we have that

$$\begin{aligned}
&\left| \int_{M \setminus B^{g_{\xi_{i_0,\alpha}}}_{R\delta_{i_0,\alpha}}(\xi_{i_0,\alpha})} f_\alpha H\left(W_{i_0,\alpha}\right)^{q_\alpha-2} \phi_\alpha \varphi_{i_0,\alpha}\, dv_g \right| \\
&\quad \le C \|W_{i_0,\alpha}\|_{2^\star}^{q_\alpha-2} \|\phi_\alpha\|_{2^\star} \left(\int_{M \setminus B^{g_{\xi_{i_0,\alpha}}}_{R\delta_{i_0,\alpha}}(\xi_{i_0,\alpha})} |\varphi_{i_0,\alpha}|^{2^\star}\, dv_g \right)^{1/2^\star} \\
&\quad \le C \|\varphi\|_{L^{2^\star}(\mathbb{R}^n \setminus B_R(0))}.
\end{aligned}$$

Since $\varphi \in L^{2^\star}(\mathbb{R}^n)$, we get that

$$\lim_{R \to +\infty} \lim_{\alpha \to +\infty} \int_{M \setminus B^{g_{\xi_{i_0,\alpha}}}_{R\delta_{i_0,\alpha}}(\xi_{i_0,\alpha})} f_\alpha H\left(W_{i_0,\alpha}\right)^{q_\alpha-2} \phi_\alpha \varphi_{i_0,\alpha}\, dv_g = 0. \qquad (83)$$

A change of variable, (63), and (64) yield

$$\begin{aligned}\int_{B^{g_{\xi_{i_0,\alpha}}}_{R\delta_{i_0,\alpha}}(\xi_{i_0,\alpha})} f_\alpha H\left(W_{i_0,\alpha}\right)^{q_\alpha-2}\phi_\alpha\varphi_{i_0,\alpha}\,dv_g &= \int_{B_R(0)} f_\alpha(\tilde{\exp}^{g_{\xi_{i_0,\alpha}}}_{\xi_{i_0,\alpha}}(\delta_{i_0,\alpha}\cdot))\\ &\times\left(\delta_{i_0,\alpha}^{2^\star-q_\alpha}\right)^{\frac{n-2}{2}} H\left(\kappa_i U_{f_0(\xi_{i_0,\alpha})}\right)^{q_\alpha-2}\tilde{\phi}_{i_0,\alpha}\varphi\,dv_{g_\alpha}\\ &= \int_{B_R(0)} U_1^{2^\star-2}\tilde{\phi}_{i_0,\alpha}\varphi\,dx + o(1),\end{aligned} \tag{84}$$

where $g_\alpha := (\tilde{\exp}^{g_{\xi_{i_0,\alpha}}}_{\xi_{i_0,\alpha}})^\star g(\delta_{i_0,\alpha}\cdot)$, and we have used that $\kappa_i = 1$ if $H = (\cdot)_+$. Moreover, it follows from Hölder's inequality that

$$\left|\int_{\mathbb{R}^n\setminus B_R(0)} U_1^{2^\star-2}\tilde{\phi}_{i_0,\alpha}\varphi\,dx\right| \leq C\|U_1\|^{2^\star-2}_{L^{2^\star}(\mathbb{R}^n\setminus B_R(0))}\|\tilde{\phi}_{i_0,\alpha}\|_{2^\star}\|\varphi\|_{2^\star}. \tag{85}$$

Plugging (83), (84), and (85) into (82) and using (79) yields

$$(\tilde{\phi}_{i_0},\varphi)_{\mathrm{Eucl}} - (2^\star-1)\int_{\mathbb{R}^n} U_1^{2^\star-2}\tilde{\phi}_{i_0}\varphi\,dx = o_\alpha(1)\,(1+\Lambda_\alpha) + \sum_j \lambda_\alpha^{i_0 j}(V_j,\varphi)_{\mathrm{Eucl}}. \tag{86}$$

It follows from (73), from (80), (33), and (43) that

$$(\tilde{\phi}_{i_0}, V_j)_{\mathrm{Eucl}} = 0 \text{ for all } j = 0,\dots,n. \tag{87}$$

Since the V_j's are solutions to (16), we then get that $\int_{\mathbb{R}^n} U_1^{2^\star-2}\tilde{\phi}_{i_0}V_j\,dx = 0$ for all $j = 0,\dots,n$. Since the V_j's are orthogonal in $D_1^2(\mathbb{R}^n)$, taking $\varphi := V_j$ in (86) yields

$$\lambda^{i_0,j} = o_\alpha(1)\,(1+\Lambda_\alpha) \text{ for all } i_0 = 1,\dots,k \text{ and } j = 0,\dots,n. \tag{88}$$

Step 3: It follows from (77) and (88) that $\Lambda_\alpha = o_\alpha(1)\,(1+\Lambda_\alpha)$, and then $\Lambda_\alpha = o_\alpha(1)$ when $\alpha \to 0$. As a consequence, (76) rewrites $\Delta_g\phi + h_0\phi = F_0''(u_0)\phi$, and then $\phi \in K_0$. Moreover, passing to the limit $\alpha \to +\infty$ in the second equation of (73) yields $\phi \in K_0^\perp$. Therefore $\phi = 0$ and then (69) rewrites

$$\phi_\alpha \rightharpoonup 0 \text{ weakly in } H_1^2(M) \text{ when } \alpha \to +\infty.$$

Similarly, (86) rewrites $\Delta_{\mathrm{Eucl}}\tilde{\phi}_{i_0} = (2^\star-1)U_1^{2^\star-2}\tilde{\phi}_{i_0}$ with $\tilde{\phi}_{i_0} \in D_1^2(\mathbb{R}^n)$. Then $\tilde{\phi}_{i_0} \in K_{BE}$ (see Subsection 2.5). On the other hand, (87) yields $\tilde{\phi}_{i_0} \in K_{BE}^\perp$. Therefore $\tilde{\phi}_{i_0} \equiv 0$, and then (79) rewrites

$$\tilde{\phi}_{i_0,\alpha} \rightharpoonup 0 \text{ weakly in } D_1^2(\mathbb{R}^n) \text{ when } \alpha \to +\infty$$

for any $i_0 = 1,\dots,k$. Since $\phi \equiv 0$, taking $\varphi := \phi_\alpha$ in (75) yields

$$\begin{aligned}\|\phi_\alpha\|^2_{h_\alpha} &= (q_\alpha-1)\int_M f_\alpha H\left(u_\alpha(z_\alpha) + \sum_{i=1}^k W_{i,\alpha}\right)^{q_\alpha-2}\phi_\alpha^2\,dv_g + o(1)\\ &\leq C\int_M |u_\alpha(z_\alpha)|^{q_\alpha-2}\phi_\alpha^2\,dv_g + C\sum_{i=1}^k\int_M |W_{i,\alpha}|^{q_\alpha-2}\phi_\alpha^2\,dv_g + o_\alpha(1),\end{aligned} \tag{89}$$

where $\lim_{\alpha\to+\infty} o_\alpha(1) = 0$. Since $\phi_\alpha \rightharpoonup 0$ when $\alpha \to +\infty$, it follows from integration theory that $\int_M |u_\alpha(z_\alpha)|^{q_\alpha-2}\phi_\alpha^2\, dv_g \to 0$ when $\alpha \to +\infty$. For any $i \in \{1,\dots,k\}$, on the one hand, for any $R > 0$, we have that

$$\int_{M\setminus B^{g_{\xi_{i,\alpha}}}_{R\delta_{i,\alpha}}(\xi_{i,\alpha})} |W_{i,\alpha}|^{q_\alpha-2}\phi_\alpha^2\, dv_g \le C\|\phi_\alpha\|^2_{2^\star}\left(\int_{M\setminus B^{g_{\xi_{i,\alpha}}}_{R\delta_{i,\alpha}}(\xi_{i,\alpha})} |W_{i,\alpha}|^{2^\star}\, dv_g\right)^{\frac{q_\alpha-2}{2^\star}}$$

$$\le C\|\phi_\alpha\|^2_{H_1^2}\left(\int_{\mathbb{R}^n\setminus B_R(0)} U_1^{2^\star}\right)^{(q_\alpha-2)/2^\star}, \tag{90}$$

and then

$$\lim_{R\to+\infty}\lim_{\alpha\to+\infty}\int_{M\setminus B^{g_{\xi_{i_0,\alpha}}}_{R\delta_{i_0,\alpha}}(\xi_{i_0,\alpha})} |W_{i,\alpha}|^{q_\alpha-2}\phi_\alpha^2\, dv_g = 0\,.$$

On the other hand, we have that

$$\int_{B^{g_{\xi_{i_0,\alpha}}}_{R\delta_{i_0,\alpha}}(\xi_{i_0,\alpha})} |W_{i,\alpha}|^{q_\alpha-2}\phi_\alpha^2\, dv_g \le C\int_{B_R(0)} U_1^{q_\alpha-2}\tilde{\phi}_{i,\alpha}\, dx\,. \tag{91}$$

Since $\tilde{\phi}_{i,\alpha} \rightharpoonup 0$ when $\alpha \to +\infty$, it follows from integration theory that the right-hand side in (91) above goes to 0 as $\alpha \to +\infty$. Plugging this latest result and (90) into (89) yields $\|\phi_\alpha\|_{h_\alpha} = o(1)$ when $\alpha \to +\infty$. A contradiction with (67). This proves (62).

We write $L_{z,(\delta_i)_i,(\xi_i)_i} := Id - \tilde{L}$, where $\tilde{L}$ is a compact operator. It then follows from (62) and Fredholm theory that $L_{z,(\delta_i)_i,(\xi_i)_i}$ is a bi-continuous isomorphism. This ends the proof of Proposition 5.2 □

5.2. Rough control of the rest

We prove the following proposition:

Proposition 5.3. *We have that*

$$R(z,(\delta_i)_i,(\xi_i)_i) \le \omega_{14}(\varepsilon,N) \tag{92}$$

for all $(z,(\delta_i)_i,(\xi_i)_i) \in B_\varepsilon(0)\times\mathcal{D}_k(\varepsilon,N)$, *where* $\lim_{\varepsilon\to0,N\to+\infty}\omega_{14}(\varepsilon,N) = 0$.

Proof of Proposition 5.3. We argue by contradiction. We assume that there exist $(q_\alpha)_\alpha \in (2,2^\star]$, $(h_\alpha)_\alpha \in L^\infty(M)$, $(f_\alpha)_\alpha \in C^0(M)$, $(z_\alpha)_\alpha \in B_1(0)$, $(u_\alpha)_\alpha \in C^1(B_1(0);H_1^2(M))$, $(G_\alpha)_\alpha \in C^{2,\theta}_{\mathrm{loc}}(H_1^2(M))$, $(\delta_{i,\alpha})_\alpha$, and $(\xi_{i,\alpha})_\alpha$ for $i = 1,\dots,k$ and $c_0 > 0$ such that

$$\lim_{\alpha\to+\infty}\|h_\alpha - h_0\|_\infty + \|f_\alpha - f_0\|_{C^0} + d_{C^{2,\theta}_B}(G_\alpha,G_0) = 0\,, \tag{93}$$

$$\lim_{\alpha\to+\infty}\delta_{i,\alpha} = 0\,,\ \lim_{\alpha\to+\infty}\delta_{i,\alpha}^{2^\star-q_\alpha} = 1\,,\ \lim_{\alpha\to0} z_\alpha = 0\,,\ \lim_{\alpha\to+\infty} q_\alpha = 2^\star\,, \tag{94}$$

$$u_\alpha(0) = u_0\,,\ \|u_\alpha\|_{C^1(B_1(0),H_1^2)} \le C_0\,,\ f_0(\xi_{i,\alpha}) \ge \nu_0 \text{ for all } i = 1,\dots,k, \tag{95}$$

$$\lim_{\alpha\to+\infty}\left(\frac{\delta_{i,\alpha}}{\delta_{j,\alpha}} + \frac{\delta_{j,\alpha}}{\delta_{i,\alpha}} + \frac{d_g(\xi_{i,\alpha},\xi_{j,\alpha})^2}{\delta_{i,\alpha}\delta_{j,\alpha}}\right) = +\infty\,, \tag{96}$$

and

$$R_\alpha := R(z_\alpha, (\delta_{i,\alpha})_i, (\xi_{i,\alpha})_i) \geq c_0 \text{ for all } \alpha \in \mathbb{N}. \tag{97}$$

We define $W_{i,\alpha} := W_{\kappa_i, \delta_{i,\alpha}, \xi_{i,\alpha}}$. In particular, Proposition 3.1 yields

$$W_{i,\alpha} \rightharpoonup 0 \text{ weakly in } H_1^2(M) \text{ when } \alpha \to +\infty. \tag{98}$$

Defining $\tilde{H}_{q_\alpha}(x) = H(x)^{q_\alpha - 2}x$ and F_α as in (71), we have that

$$\begin{aligned}
R_\alpha &= \left\| u_\alpha(z_\alpha) + \sum_{i=1}^k W_{i,\alpha} - (\Delta_g + h_\alpha)^{-1}\left(F'_\alpha\left(u_\alpha(z_\alpha) + \sum_{i=1}^k W_{i,\alpha}\right)\right)\right\|_{H_1^2} \\
&\leq \left\| u_\alpha(z_\alpha) - (\Delta_g + h_\alpha)^{-1}\left(F'_\alpha(u_\alpha(z_\alpha))\right)\right\|_{H_1^2} \\
&\quad + \sum_{i=1}^k \left\| W_{i,\alpha} - (\Delta_g + h_\alpha)^{-1}(f_\alpha \tilde{H}_{q_\alpha}(W_{i,\alpha}))\right\|_{H_1^2} \\
&\quad + \left\| (\Delta_g + h_\alpha)^{-1}\left(F'_\alpha\left(u_\alpha(z_\alpha) + \sum_{i=1}^k W_{i,\alpha}\right) - F'_\alpha(u_\alpha(z_\alpha))\right)\right. \\
&\quad \left. - \sum_{i=1}^k (\Delta_g + h_\alpha)^{-1}(f_\alpha \tilde{H}_{q_\alpha}(W_{i,\alpha}))\right\|_{H_1^2}.
\end{aligned}$$

The control (95) yields $\lim_{\alpha\to+\infty} u_\alpha(z_\alpha) = u_0$ in $H_1^2(M)$. The convergence (93) then yields

$$\begin{aligned}
&\lim_{\alpha\to+\infty} \left\| u_\alpha(z_\alpha) - (\Delta_g + h_\alpha)^{-1}(F'_\alpha(u_\alpha(z_\alpha)))\right\|_{H_1^2} \\
&\qquad = \left\| u_0 - (\Delta_g + h_0)^{-1}(F'_0(u_0))\right\|_{H_1^2} = 0\,.
\end{aligned}$$

Since $(u_\alpha(z_\alpha) + \sum_{i=1}^k W_{i,\alpha})_\alpha \rightharpoonup u_0$ weakly in $H_1^2(M)$, it follows from the convergence (93) of $(G_\alpha)_\alpha$ and the Definition 2.1 of subcriticality of G_0 that

$$G'_\alpha\left(u_\alpha(z_\alpha) + \sum_{i=1}^k W_{i,\alpha}\right) \to G'_0(u_0) \text{ strongly in } H_1^2(M) \text{ when } \alpha \to +\infty. \tag{99}$$

As a consequence, we get with the Riesz correspondence (23) that

$$R_\alpha \leq o(1) + \sum_{i=1}^k \left\| W_{i,\alpha} - (\Delta_g + h_\alpha)^{-1}(f_\alpha \tilde{H}_{q_\alpha}(W_{i,\alpha})\right\|_{H_1^2} + C\|A_\alpha\|_{\frac{2n}{n+2}},$$

where

$$A_\alpha := \tilde{H}_{q_\alpha}\left(u_\alpha(z_\alpha) + \sum_{i=1}^k W_{i,\alpha}\right) - \tilde{H}_{q_\alpha}(u_\alpha(z_\alpha)) - \sum_{i=1}^k \tilde{H}_{q_\alpha}(W_{i,\alpha}).$$

As easily checked, for all family $(X_i)_{i=0,\dots,k} \in \mathbb{R}$, we have that

$$\left|\tilde{H}_{q_\alpha}\left(\sum_{i=0}^k X_i\right) - \sum_{i=0}^k \tilde{H}_{q_\alpha}(X_i)\right| \leq C \sum_{i\neq j} |X_i| \cdot |X_j|^{q_\alpha - 2}$$

for all $\alpha \in \mathbb{N}$ large. Therefore, with Hölder's inequality, we get that

$$\begin{aligned}\|A_\alpha\|_{\frac{2n}{n+2}}^{\frac{2n}{n+2}} &\le C\sum_i \int_M |u_\alpha(z_\alpha)|^{\frac{2n}{n+2}} |W_{i,\alpha}|^{(q_\alpha-2)\frac{2n}{n+2}}\, dv_g \\ &+ C\sum_i \int_M |u_\alpha(z_\alpha)|^{(q_\alpha-2)\frac{2n}{n+2}} |W_{i,\alpha}|^{\frac{2n}{n+2}}\, dv_g \\ &+ C\sum_{i\neq j} \int_M |W_{i,\alpha}|^{\frac{2n}{n+2}} |W_{j,\alpha}|^{(q_\alpha-2)\frac{2n}{n+2}}\, dv_g\,.\end{aligned}$$

Since $u_\alpha(z_\alpha) \to u_0$ in $L^{2^\star}(M)$, $(|W_{i,\alpha}|^{(q_\alpha-2)\frac{2n}{n+2}})_\alpha$ is bounded in $L^{(n+2)/4}(M)$ and goes to zero a.e. on M when $\alpha \to +\infty$, integration theory yields the convergence to 0 of the first term of the right-hand side when $\alpha \to +\infty$. Similarly, the second term goes to 0 as $\alpha \to +\infty$. The expression (25), the property (13), and Proposition 3.4 yield the convergence to 0 of the third term when $\alpha \to +\infty$. Therefore, we get that $(\|A_\alpha\|_{\frac{2n}{n+2}})_\alpha \to 0$, and then

$$R_\alpha \le o(1) + \sum_{i=1}^k \left\| W_{i,\alpha} - (\Delta_g + h_\alpha)^{-1}\left(f_\alpha \tilde{H}_{q_\alpha}(W_{i,\alpha})\right)\right\|_{H_1^2}\,.$$

We define

$$W_{i,\alpha}^0 := \chi(d_{g_{\xi_{i,\alpha}}}(\cdot,\xi_{i,\alpha}))\Lambda_{\xi_{i,\alpha}}\left(\frac{\delta_{i,\alpha}\sqrt{\frac{n(n-2)}{f_0(\xi_{i,\alpha})}}}{\delta_{i,\alpha}^2 + d_{g_{\xi_{i,\alpha}}}(\cdot,\xi_{i,\alpha})^2}\right)^{\frac{n-2}{2}}$$

so that $W_{i,\alpha} = \kappa_i W_{i,\alpha}^0 + o(1)$ when $\alpha \to +\infty$ (see (12)). Therefore, since $\kappa_i = 1$ if $H = (\cdot)_+$, (23) yields

$$\begin{aligned}R_\alpha &\le o(1) + \sum_{i=1}^k \left\|\kappa_i W_{i,\alpha}^0 - (\Delta_g + h_\alpha)^{-1}(f_\alpha \tilde{H}_{q_\alpha}(\kappa_i W_{i,\alpha}^0)\right\|_{H_1^2} \\ &\le o(1) + \sum_{i=1}^k \|(\Delta_g + h_\alpha)W_{i,\alpha}^0 - f_\alpha (W_{i,\alpha}^0)^{q_\alpha-1}\|_{\frac{2n}{n+2}}\,.\end{aligned} \tag{100}$$

In the sequel, $o(1)_{\frac{2n}{n+2}}$ denotes a function going to 0 in $L^{\frac{2n}{n+2}}(M)$ when $\alpha \to +\infty$. We define $c_n := \frac{n-2}{4(n-1)}$, and we let R_g be the scalar curvature of g. We denote $L_g := \Delta_g + c_n R_g$ the conformal Laplacian. If $g' := \varpi^{2^\star-2}g$ and $\varpi \in C^2(M)$ positive, the conformal invariance properties of L_g yields

$$L_g'(\varphi) := \varpi^{1-2^\star} L_g(\varpi\varphi)$$

for all $\varphi \in C^2(M)$. Using the expression of the Laplacian in radial coordinates, omitting the index i and writing $r := d_{g_\xi}(x,\xi)$, we get that

$$\begin{aligned}(\Delta_g + h_\alpha)W_{i,\alpha}^0 &= L_g W_{i,\alpha}^0 + (h_\alpha - c_n R_g)W_{i,\alpha}^0 \\ &= L_{\Lambda_\xi^{2-2^\star} g_\xi} W_{i,\alpha}^0 + o(1)_{\frac{2n}{n+2}} = \Lambda_\xi^{2^\star-1} L_{g_\xi}(\Lambda_\xi^{-1} W_{i,\alpha}^0) + o(1)_{\frac{2n}{n+2}}\end{aligned}$$

$$
\begin{aligned}
&= \Lambda_\xi^{2^\star-1}\Delta_{g_\xi}\left(\chi(r)\delta^{-\frac{n-2}{2}}U_{f_0(\xi)}(\delta^{-1}r)\right)+o(1)_{\frac{2n}{n+2}}\\
&= \Lambda_\xi^{2^\star-1}\Delta_{\mathrm{Eucl}}\left(\chi(r)\delta^{-\frac{n-2}{2}}U_{f_0(\xi)}(\delta^{-1}r)\right)\\
&- \partial_r \ln\sqrt{|g_\xi|}\partial_r\left(\chi(r)\delta^{-\frac{n-2}{2}}U_{f_0(\xi)}(\delta^{-1}r)\right)+o(1)_{\frac{2n}{n+2}}\\
&= \Lambda_\xi^{2^\star-1}\Delta_{\mathrm{Eucl}}\left(\chi(r)\delta^{-\frac{n-2}{2}}U_{f_0(\xi)}(\delta^{-1}r)\right)\\
&\quad +O\left(\frac{\delta^{\frac{n-2}{2}}}{(\delta^2+r^2)^{\frac{n-2}{2}}}\right)+o(1)_{\frac{2n}{n+2}}\\
&= \Lambda_\xi^{2^\star-1}\chi(r)\Delta_{\mathrm{Eucl}}\left(\delta^{-\frac{n-2}{2}}U_{f_0(\xi)}(\delta^{-1}r)\right)+o(1)_{\frac{2n}{n+2}}\\
&= \Lambda_\xi^{2^\star-1}\chi(r)f_0(\xi)\left(\delta^{-\frac{n-2}{2}}U_{f_0(\xi)}(\delta^{-1}r)\right)^{2^\star-1}+o(1)_{\frac{2n}{n+2}}\\
&= f_0(\xi)\left(\chi(r)\Lambda_\xi\delta^{-\frac{n-2}{2}}U_{f_0(\xi)}(\delta^{-1}r)\right)^{2^\star-1}+o(1)_{\frac{2n}{n+2}}\\
&= f_0(\xi)(W_{i,\alpha}^0)^{2^\star-1}+o(1)_{\frac{2n}{n+2}}\\
&= f_0(W_{i,\alpha}^0)^{2^\star-1}+o(1)_{\frac{2n}{n+2}} = f_\alpha(W_{i,\alpha}^0)^{2^\star-1}+o(1)_{\frac{2n}{n+2}}\,.
\end{aligned}
$$

Therefore, it follows from (100) that

$$
R_\alpha \le o(1) + C\sum_{i=1}^k \|(W_{i,\alpha}^0)^{2^\star-1}-(W_{i,\alpha}^0)^{q_\alpha-1}\|_{\frac{2n}{n+2}}\,. \tag{101}
$$

We fix $i\in\{1,\dots,k\}$. For any $R>0$, a change of variable and (94) yields

$$
\begin{aligned}
&\int_{B_{R\delta_{i,\alpha}}^{g_{\xi_{i,\alpha}}}(\xi_{i,\alpha})}\left|(W_{i,\alpha}^0)^{2^\star-1}-(W_{i,\alpha}^0)^{q_\alpha-1}\right|^{\frac{2n}{n+2}}dv_g\\
&\le C\int_{B_{R\delta_{i,\alpha}}^{g_{\xi_{i,\alpha}}}(\xi_{i,\alpha})}\left|(W_{i,\alpha}^0)^{2^\star-1}-(W_{i,\alpha}^0)^{q_\alpha-1}\right|^{\frac{2n}{n+2}}dv_{g_{\xi_{i,\alpha}}} \qquad (102)\\
&\le C\int_{B_R(0)}\left|U_{f_0(\xi_{i,\alpha})}^{2^\star-1}-\delta_{i,\alpha}^{\frac{n-2}{2}(2^\star-q_\alpha)}U_{f_0(\xi_{i,\alpha})}^{q_\alpha-1}\right|^{\frac{2n}{n+2}}dx=o(1)
\end{aligned}
$$

when $\alpha\to+\infty$. Independently, we have that

$$
\begin{aligned}
&\int_{M\setminus B_{R\delta_{i,\alpha}}^{g_{\xi_{i,\alpha}}}(\xi_{i,\alpha})}\left|(W_{i,\alpha}^0)^{2^\star-1}-(W_{i,\alpha}^0)^{q_\alpha-1}\right|^{\frac{2n}{n+2}}dv_g\\
&\le C\int_{M\setminus B_{R\delta_{i,\alpha}}^{g_{\xi_{i,\alpha}}}(\xi_{i,\alpha})}\left|W_{i,\alpha}^0\right|^{2^\star}dv_g+C\left(\int_{M\setminus B_{R\delta_{i,\alpha}}^{g_{\xi_{i,\alpha}}}(\xi_{i,\alpha})}\left|W_{i,\alpha}^0\right|^{2^\star}dv_g\right)^{\frac{q_\alpha-1}{2^\star-1}}\\
&\le C\int_{\mathbb{R}^n\setminus B_R(0)}U_1^{2^\star}dx+C\left(\int_{\mathbb{R}^n\setminus B_R(0)}U_1^{2^\star}dx\right)^{\frac{q_\alpha-1}{2^\star-1}}.
\end{aligned}
$$

Then

$$\lim_{R\to+\infty}\lim_{\alpha\to+\infty}\int_{M\setminus B^{g_{\xi_{i,\alpha}}}_{R\delta_{i,\alpha}}(\xi_{i,\alpha})}\left|(W^0_{i,\alpha})^{2^\star-1}-(W^0_{i,\alpha})^{q_\alpha-1}\right|^{\frac{2n}{n+2}}dv_g=0\,. \tag{103}$$

Plugging (102) and (103) into (101) yields $R_\alpha=o(1)$ when $\alpha\to+\infty$, a contradiction with (97). This proves Proposition 5.3. □

5.3. Proof of Proposition 5.1 via a fixed-point argument

We let $\varepsilon, N>0$ satisfy the hypotheses of Proposition 5.2 to be fixed later, and we let h,f,G,u satisfy the hypotheses of Theorem 1.1. We consider $(z,(\delta_i)_i,(\xi_i)_i)\in B_\varepsilon(0)\times\mathcal{D}_k(\varepsilon,N)$, and we define $K:=K_{(\delta_i)_i,(\xi_i)_i}$. For any $\phi\in K^\perp\subset H_1^2(M)$, we have that

$$\Pi_{K^\perp}\left(u(z)+\sum_{i=1}^k W_{\kappa_i,\delta_i,\xi_i}+\phi-(\Delta_g+h)^{-1}\left(F'\left(u(z)+\sum_{i=1}^k W_{\kappa_i,\delta_i,\xi_i}+\phi\right)\right)\right)=0 \tag{104}$$

if and only if

$$\phi=T(\phi),$$

where $T:K^\perp\to K^\perp$ is such that

$$T(\phi):=L^{-1}\circ\Pi_{K^\perp}\circ(\Delta_g+h)^{-1}(N(\phi))-L^{-1}\circ\Pi_{K^\perp}(R),$$

where $L:=L_{(\delta_i)_i,(\xi_i)_i}$,

$$\begin{aligned}N(\phi):=&\,F'\left(u(z)+\sum_{i=1}^k W_{\kappa_i,\delta_i,\xi_i}+\phi\right)-F'\left(u(z)+\sum_{i=1}^k W_{\kappa_i,\delta_i,\xi_i}\right)\\&-F''\left(u(z)+\sum_{i=1}^k W_{\kappa_i,\delta_i,\xi_i}\right)\phi\end{aligned}$$

and

$$R:=u(z)+\sum_{i=1}^k W_{\kappa_i,\delta_i,\xi_i}-(\Delta_g+h)^{-1}\left(F'\left(u(z)+\sum_{i=1}^k W_{\kappa_i,\delta_i,\xi_i}\right)\right).$$

We prove the existence of a solution to (104) via Picard's Fixed Point Theorem. We let $\phi_1,\phi_2\in K^\perp$ be two test-functions. Since $\Pi_{K^\perp}:H_1^2(M)\to H_1^2(M)$ is 1-Lipschitz continuous, it follows from (62) that

$$\begin{aligned}&\|T(\phi_1)-T(\phi_2)\|_{H_1^2}\le C\|N(\phi_1)-N(\phi_2)\|_{H_1^2}\\&\le C\Bigg\|F'\left(u(z)+\sum_{i=1}^k W_{\kappa_i,\delta_i,\xi_i}+\phi_1\right)-F'\left(u(z)+\sum_{i=1}^k W_{\kappa_i,\delta_i,\xi_i}+\phi_2\right)\\&\quad-F''\left(u(z)+\sum_{i=1}^k W_{\kappa_i,\delta_i,\xi_i}\right)(\phi_1-\phi_2)\Bigg\|_{(H_1^2)'}.\end{aligned}$$

It then follows from the mean value inequality that

$$\|T(\phi_1) - T(\phi_2)\|_{H_1^2(M)} \leq C \cdot S \cdot \|\phi_1 - \phi_2\|_{H_1^2}, \tag{105}$$

where

$$\begin{aligned} S := & \sup_{t\in[0,1]} \left\| F''\left(u(z) + \sum_{i=1}^{k} W_{\kappa_i,\delta_i,\xi_i} + \phi_1 + t(\phi_2 - \phi_1)\right) \right. \\ & \left. - F''\left(u(z) + \sum_{i=1}^{k} W_{\kappa_i,\delta_i,\xi_i}\right) \right\|_{H_1^2 \to (H_1^2)'} \\ \leq & \sup_{t\in[0,1]} \left(\|F\|_{C^{2,\theta}(B_{\tilde R}(0))} \cdot \|\phi_1 + t(\phi_2 - \phi_1)\|_{H_1^2}^{\theta} \right) \\ \leq & C\|F\|_{C^{2,\theta}(B_{\tilde R}(0))} \cdot \left(\|\phi_1\|_{H_1^2}^{\theta} + \|\phi_2\|_{H_1^2}^{\theta} \right), \end{aligned} \tag{106}$$

with $\tilde R = \tilde R(z, (\delta_i)_i, (\xi_i)_i) := \|u(z)\|_{H_1^2} + \sum_{i=1} \|W_{\kappa_i,\delta_i,\xi_i}\|_{H_1^2} + 1$, $\|\phi_1\|_{H_1^2}, \|\phi_1\|_{H_1^2} \leq 1$. It then follows from (6) and Proposition 3.1 that there holds $\tilde R(z, (\delta_i)_i, (\xi_i)_i) \leq C$. As easily checked, for $2 < q \leq 2^\star$, we have that

$$F''(v)(\psi_1, \psi_2) = (q-1) \int_M f H(v)^{q-2} \psi_1 \psi_2 \, dv_g + G''(v)(\psi_1, \psi_2)$$

for all $v \in H_1^2(M)$ and all $\psi_1, \psi_2 \in H_1^2(M)$. Without loss of generality, we may assume that $0 < \theta < 2^\star - 2$. Requiring that $\varepsilon < 1$ and using (6), we then get that

$$\|F\|_{C^{2,\theta}(B_{\tilde R}(0))} \leq C(\tilde R, \theta) \tag{107}$$

for $\epsilon > 0$ small enough. Plugging together (105), (106), and (107) yields

$$\|T(\phi_1) - T(\phi_2)\|_{H_1^2(M)} \leq C_1 \cdot \left(\|\phi_1\|_{H_1^2}^{\theta} + \|\phi_2\|_{H_1^2}^{\theta} \right) \cdot \|\phi_1 - \phi_2\|_{H_1^2} \tag{108}$$

for $\phi_1, \phi_2 \in K^\perp$ such that $\|\phi_1\|_{H_1^2}, \|\phi_1\|_{H_1^2} \leq 1$. Moreover, it follows from (23) and (5) that

$$\|T(0)\|_{H_1^2} \leq C\|R\|_{H_1^2} \leq C_2 R(z, (\delta_i)_i, (\xi_i)_i). \tag{109}$$

We define

$$c := 2C_2 R(z, (\delta_i)_i, (\xi_i)_i).$$

We let $\phi_1, \phi_2 \in K^\perp \cap \overline{B}_c(0)$: it then follows from (108) and (109) that

$$\|T(\phi_1) - T(\phi_2)\|_{H_1^2(M)} \leq 2C_1(2C_2)^\theta R(z, (\delta_i)_i, (\xi_i)_i)^\theta \cdot \|\phi_1 - \phi_2\|_{H_1^2}$$

and

$$\begin{aligned} \|T(\phi_1)\|_{H_1^2} & \leq C_2 R(z, (\delta_i)_i, (\xi_i)_i) + C_1(2C_2)^{1+\theta} R(z, (\delta_i)_i, (\xi_i)_i)^{1+\theta} \\ & \leq \left(C_2 + C_2(2C_2)^{1+\theta} R(z, (\delta_i)_i, (\xi_i)_i)^\theta\right) \cdot R(z, (\delta_i)_i, (\xi_i)_i). \end{aligned}$$

It follows from Proposition 5.3, that there exists $\varepsilon > 0$ and $N > 0$ such that

$$R(z, (\delta_i)_i, (\xi_i)_i)^\theta \leq \min\left\{ \frac{1}{C_1 C_2^\theta 2^{1+\theta}} \, ; \, \frac{1}{4C_2(2C_2)^\theta} \right\}$$

for all $z \in B_\varepsilon(0)$ and $((\delta_i)_i, (\xi_i)_i) \in \mathcal{D}_k(\varepsilon, N)$. It then follows that for such a choice, the map T is 1/2-Lipschitz from $\overline{B}_c(0)$ onto itself. It then follows from Picard's fixed point theorem that there exists a unique solution $\phi(z, (\delta_i)_i, (\xi_i)_i) \in \overline{B}_c(0) \cap K^\perp$ to $T(\phi(z, (\delta_i)_i, (\xi_i)_i)) = \phi(z, (\delta_i)_i, (\xi_i)_i)$, in particular

$$\|\phi(z, (\delta_i)_i, (\xi_i)_i)\|_{H_1^2} \leq 2C_2 R(z, (\delta_i)_i, (\xi_i)_i).$$

We are left with proving the C^1-regularity of ϕ. We define the map

$$\begin{array}{rccc} \mathcal{F}: & B_\varepsilon(0) \times \mathcal{D}_k(\varepsilon, N) \times H_1^2(M) & \to & H_1^2(M) \\ & (z, (\delta_i)_i, (\xi_i)_i, \phi) & \mapsto & \mathcal{F}(z, (\delta_i)_i, (\xi_i)_i, \phi), \end{array}$$

where

$$\begin{aligned} \mathcal{F}(z, (\delta_i)_i, (\xi_i)_i, \phi) := \Pi_K(\phi) + \Pi_{K^\perp}\Big(u(z) + \sum\nolimits_{i=1}^k W_{\kappa_i, \delta_i, \xi_i} + \Pi_{K^\perp}(\phi) \\ - (\Delta_g + h)^{-1}(F'(u(z) + \sum\nolimits_{i=1}^k W_{\kappa_i, \delta_i, \xi_i} + \Pi_{K^\perp}(\phi)))\Big). \end{aligned}$$

By Proposition 5.2, we get that the differential with respect to ϕ is an isomorphism of $H_1^2(M)$ for all $(z, (\delta_i)_i, (\xi_i)_i, \phi) \in B_\varepsilon(0) \times \mathcal{D}_k(\varepsilon, N) \times H_1^2(M)$, with $\|\phi\|_{H_1^2} < c_0$ for some $c_0 > 0$ small. Since $\mathcal{F}(z, (\delta_i)_i, (\xi_i)_i, \phi(z, (\delta_i)_i, (\xi_i)_i)) = 0$ for all $(z, (\delta_i)_i, (\xi_i)_i) \in B_\varepsilon(0) \times \mathcal{D}_k(\varepsilon, N)$, it follows from the implicit functions theorem that $(z, (\delta_i)_i, (\xi_i)_i) \mapsto \phi(z, (\delta_i)_i, (\xi_i)_i)$ is C^1 on $B_\varepsilon(0) \times \mathcal{D}_k(\varepsilon, N)$. This ends the proof of Proposition 5.1. □

6. Equivalence of the critical points

We prove Theorem 1.1 in this section. With Proposition 5.1 above, this amounts to prove the equivalence of the critical points for $\varepsilon > 0$ small and $N > 0$ large. For $\varepsilon, N > 0$ satisfying the hypothesis of Proposition 5.1, there exists $\phi \in C^1(B_\varepsilon(0) \times \mathcal{D}_k(\varepsilon, N), H_1^2(M)$ such that

$$\Pi_{K^\perp_{(\delta_i)_i, (\xi_i)_i}}\left(u(z, (\delta_i)_i, (\xi_i)_i) - (\Delta_g + h)^{-1}(F'(u(z, (\delta_i)_i, (\xi_i)_i)))\right) = 0, \quad (110)$$

where

$$u(z, (\delta_i)_i, (\xi_i)_i) := u(z) + \sum_{i=1}^k W_{\kappa_i, \delta_i, \xi_i} + \phi(z, (\delta_i)_i, (\xi_i)_i),$$

and

$$\phi(z, (\delta_i)_i, (\xi_i)_i) \in K^\perp_{(\delta_i)_i, (\xi_i)_i} \text{ and } \|\phi(z, (\delta_i)_i, (\xi_i)_i)\|_{H_1^2} \leq C \cdot R(z, (\delta_i)_i, (\xi_i)_i) \quad (111)$$

for $(z, (\delta_i)_i, (\xi_i)_i) \in \mathcal{D}_k(\varepsilon, N)$.

It then follows from (59) that there exist $\lambda^{ij}(z,(\delta_i)_i,(\xi_i)_i) \in \mathbb{R}$ ($i = 1,\dots,k$ and $j = 0,\dots,n$) and $\mu^l(z,(\delta_i)_i,(\xi_i)_i) \in \mathbb{R}$ ($l = 1,\dots,d$) such that

$$\begin{aligned} &\Pi_{K_{(\delta_i)_i,(\xi_i)_i}}\left(u(z,(\delta_i)_i,(\xi_i)_i) - (\Delta_g + h)^{-1}(F'(u(z,(\delta_i)_i,(\xi_i)_i)))\right) \\ &= \sum_{i=1}^{k}\sum_{j=0}^{n} \lambda^{ij}(z,(\delta_i)_i,(\xi_i)_i) Z_{\delta_i,\xi_i,j} + \sum_{l=1}^{d} \mu^l(z,(\delta_i)_i,(\xi_i)_i)\varphi_l\,. \end{aligned} \tag{112}$$

It then follows from (110) and (112) that for any $\varphi \in H_1^2(M)$, we have that

$$\begin{aligned} DJ(u(z,(\delta_i)_i,(\xi_i)_i))\varphi &= \left(u(z,(\delta_i)_i,(\xi_i)_i) - (\Delta_g + h)^{-1}(F'(u(z,(\delta_i)_i,(\xi_i)_i))),\varphi\right)_h \\ &= \sum_{i=1}^{k}\sum_{j=0}^{n} \lambda^{ij}(z,(\delta_i)_i,(\xi_i)_i)\left(Z_{\delta_i,\xi_i,j},\varphi\right)_h + \sum_{l=1}^{d} \mu^l(z,(\delta_i)_i,(\xi_i)_i)(\varphi_l,\varphi)_h\,. \end{aligned} \tag{113}$$

If $u(z,(\delta_i)_i,(\xi_i)_i)$ is a critical point of J, then $(z,(\delta_i)_i,(\xi_i)_i) \in B_\varepsilon(0) \times \mathcal{D}_k(\varepsilon,N)$ is a critical point for $(z,(\delta_i)_i,(\xi_i)_i) \mapsto J(u(z,(\delta_i)_i,(\xi_i)_i))$. Conversely, we assume that $(z,(\delta_i)_i,(\xi_i)_i) \in B_\varepsilon(0) \times \mathcal{D}_k(\varepsilon,N)$ is a critical point for the map $(z,(\delta_i)_i,(\xi_i)_i) \mapsto J(u(z,(\delta_i)_i,(\xi_i)_i))$. We then get that

$$\begin{aligned} 0 &= \frac{\partial}{\partial z_{l_0}} J(u(z,(\delta_i)_i,(\xi_i)_i)) \\ &= DJ(u(z,(\delta_i)_i,(\xi_i)_i)).\left(\partial_{l_0}u(z) + \partial_{z_{l_0}}\phi(z,(\delta_i)_i,(\xi_i)_i)\right), \end{aligned} \tag{114}$$

$$\begin{aligned} 0 &= \frac{\partial}{\partial \delta_{i_0}} J(u(z,(\delta_i)_i,(\xi_i)_i)) \\ &= DJ(u(z,(\delta_i)_i,(\xi_i)_i)).\left(\partial_{\delta_{i_0}} W_{\alpha_{i_0},\delta_{i_0},\xi_{i_0}} + \partial_{\delta_{i_0}}\phi(z,(\delta_i)_i,(\xi_i)_i)\right), \end{aligned} \tag{115}$$

$$\begin{aligned} 0 &= \frac{\partial}{\partial (\xi_{i_0})_{j_0}} J(u(z,(\delta_i)_i,(\xi_i)_i)) \\ &= DJ(u(z,(\delta_i)_i,(\xi_i)_i)).(\partial_{(\xi_{i_0})_{j_0}} W_{\alpha_{i_0},\delta_{i_0},\xi_{i_0}} + \partial_{(\xi_{i_0})_{j_0}}\phi(z,(\delta_i)_i,(\xi_i)_i)) \end{aligned} \tag{116}$$

for all $l_0 = 1,\dots,d$, $i_0 = 1,\dots,k$, and $j_0 = 0,\dots,n$. From now on, for the sake of clearness, we omit the variables $(z,(\delta_i)_i,(\xi_i)_i)$. We define

$$\Lambda := \sum_{i=1}^{k}\sum_{j=0}^{n} |\lambda^{ij}| + \sum_{l=1}^{d} |\mu^l|\,.$$

We are going to prove that $\Lambda = 0$, which will imply that $u(z,(\delta_i)_i,(\xi_i)_i)$ is a critical point of J due to (113).

6.1. Consequences of (114)

It follows from (113) and (114) that

$$\begin{aligned} &\sum\nolimits_{i=1}^{k}\sum\nolimits_{j=0}^{n} \lambda^{ij}\left((Z_{\delta_i,\xi_i,j},\partial_{l_0}u)_h + (Z_{\delta_i,\xi_i,j},\partial_{z_{l_0}}\phi)_h\right) \\ &\quad + \sum\nolimits_{l=1}^{d} \mu^l\left((\varphi_l,\partial_{l_0}u)_h + (\varphi_l,\partial_{z_{l_0}}\phi)_h\right) = 0\,. \end{aligned} \tag{117}$$

It follows from (111) that

$$(Z_{\delta_i,\xi_i,j},\phi)_h=(\varphi_l,\phi)_h=0\,. \tag{118}$$

Differentiating (118) with respect to z_{l_0}, we get that

$$(Z_{\delta_i,\xi_i,j},\partial_{z_{l_0}}\phi)_h=(\varphi_l,\partial_{z_{l_0}}\phi)_h=0,$$

and therefore (117) rewrites

$$\sum_{i=1}^{k}\sum_{j=0}^{n}\lambda^{ij}(Z_{\delta_i,\xi_i,j},\partial_{l_0}u)_h+\sum_{l=1}^{d}\mu^l(\varphi_l,\partial_{l_0}u)_h=0\,, \tag{119}$$

and therefore, since $\|h-h_0\|_\infty<\varepsilon$, for all $l_0=1,\dots,d$, (25) and (26) yield

$$\left|\sum_{l=1}^{d}\mu^l\left(\varphi_l,\frac{\Pi_{K_0}^{h_0}(\partial_{l_0}u)}{\|\partial_{l_0}u\|_{H_1^2}}\right)_{h_0}\right|\le C\cdot\varepsilon\cdot\Lambda+C\sup_{i,j}|\lambda^{ij}|\,, \tag{120}$$

where $\Pi_{K_0}^{h_0}$ is the orthogonal projection on K_0 (see (3) and (4)) with respect to the Hilbert structure $(\cdot,\cdot)_{h_0}$. We define the matrix $(A(z))_{ll'}:=(\varphi_l,\Pi_{K_0}^{h_0}(\partial_{l'}u))_{h_0}$ for all $l,l'\in\{1,\dots,d\}$. With no loss of generality, we can assume that the basis β_0 is $\{\varphi_1,\dots,\varphi_d\}$: it then follows from (8) and Cramer's explicit formula that the coefficients of the inverse of the matrix $A(z)$ are bounded from above by a constant C. Therefore, it follows from (120) that

$$\sum_{l=1}^{d}|\mu^l|\le C\cdot\varepsilon\cdot\Lambda+C\sup_{i,j}|\lambda^{ij}| \tag{121}$$

for $(z,(\delta_i)_i,(\xi_i)_i)\in B_\varepsilon(0)\times\mathcal{D}_k(\varepsilon,N)$.

6.2. Consequences of (115)

It follows from (113) and (115) that

$$\begin{aligned}\sum_{i=1}^{k}\sum_{j=0}^{n}\lambda^{ij}\left((Z_{\delta_i,\xi_i,j},\partial_{\delta_{i_0}}W_{\delta_{i_0},\xi_{i_0}})_h+(Z_{\delta_i,\xi_i,j},\partial_{\delta_{i_0}}\phi)_h\right)&\\ +\sum_{l=1}^{d}\mu^l\left((\varphi_l,\partial_{\delta_{i_0}}W_{\delta_{i_0},\xi_{i_0}})_h+(\varphi_l,\partial_{\delta_{i_0}}\phi)_h\right)&=0\,.\end{aligned} \tag{122}$$

Differentiating (118) with respect to δ_{i_0}, we get that $(\varphi_l,\partial_{\delta_{i_0}}\phi)_h=0$ and also $(\partial_{\delta_{i_0}}Z_{\delta_i,\xi_i,j},\phi)_h+(Z_{\delta_i,\xi_i,j},\partial_{\delta_{i_0}}\phi)_h=0$, and therefore (122) rewrites

$$\begin{aligned}\left|\sum_{i=1}^{k}\sum_{j=0}^{n}\lambda^{ij}\left(Z_{\delta_i,\xi_i,j},\partial_{\delta_{i_0}}W_{\delta_{i_0},\xi_{i_0}}\right)_h\right|&\le\sum_{l=1}^{d}|\mu^l|\cdot|(\varphi_l,\partial_{\delta_{i_0}}W_{\delta_{i_0},\xi_{i_0}})_h|\\ &\quad+\sum_{i=1}^{k}\sum_{j=0}^{n}|\lambda^{ij}|\cdot\|\partial_{\delta_{i_0}}Z_{\delta_i,\xi_i,j}\|_{H_1^2}\|\phi\|_{H_1^2}.\end{aligned} \tag{123}$$

For any $i = 1, \dots, k$ and $j = 0, \dots, n$, by (50) and Corollary 3.3, we get that

$$\begin{aligned}
&\left(Z_{\delta_i,\xi_i,j}, \partial_{\delta_{i_0}} W_{\kappa_{i_0}\delta_{i_0},\xi_{i_0}}\right)_h \\
&= \kappa_{i_0} \frac{n-2}{2} \left(\frac{n(n-2)}{f_0(\xi_{i_0})}\right)^{\frac{n-2}{4}} \cdot \frac{1}{\delta_{i_0}} \cdot \left((Z_{\delta_i,\xi_i,j}, Z_{\delta_{i_0},\xi_{i_0},0})_h + o(1)\right) \\
&= \kappa_{i_0} \frac{n-2}{2} \left(\frac{n(n-2)}{f_0(\xi_{i_0})}\right)^{\frac{n-2}{4}} \cdot \frac{1}{\delta_{i_0}} \cdot \left(\delta_{i,i_0}\delta_{j,0}\|\nabla V_0\|_2 + o(1)\right), \qquad (124)
\end{aligned}$$

where $|o(1)| \le \omega_{15}(\varepsilon, N)$ and $\lim_{\varepsilon \to 0, N \to +\infty} \omega_{15}(\varepsilon, N) = 0$. Plugging (124) into (123) and using (50) yield

$$\begin{aligned}
&|\lambda^{i_0,0}| \cdot \frac{n-2}{2} \left(\frac{n(n-2)}{f_0(\xi_{i_0})}\right)^{\frac{n-2}{4}} \cdot \frac{1}{\delta_{i_0}} \|\nabla V_0\|_2 \\
&\le \Bigg(\sum_{l=1}^{d} \frac{n-2}{2} \left(\frac{n(n-2)}{f_0(\xi_{i_0})}\right)^{\frac{n-2}{4}} \frac{1}{\delta_{i_0}} \left|(\varphi_l, Z_{\delta_{i_0},\xi_{i_0},0})_h\right| \\
&\quad + \sum_{j=0}^{n} \|\partial_{\delta_{i_0}} Z_{\delta_{i_0},\xi_{i_0},j}\|_{H_1^2} \|\phi\|_{H_1^2} + \delta_{i_0}^{-1} \omega_{12}(\varepsilon, N) \Bigg) \cdot \Lambda .
\end{aligned}$$

It then follows from (52), (111), Proposition 3.1, and the expression (25) that

$$|\lambda^{i_0,0}| \le \omega_{16}(\varepsilon, N) \cdot \Lambda \qquad (125)$$

for all $i_0 = 1, \dots, k$, where $\lim_{\varepsilon \to 0, N \to +\infty} \omega_{16}(\varepsilon, N) = 0$.

6.3. Conclusion for the equivalence

Arguing as above for (116), we get that

$$|\lambda^{i_0,j}| \le \omega_{17}(\varepsilon, N) \cdot \Lambda \qquad (126)$$

for all $i_0 = 1, \dots, k$ and $j = 1, \dots, n$, where $\lim_{\varepsilon \to 0, N \to +\infty} \omega_{17}(\varepsilon, N) = 0$. Summing (121), (125), and (126) yields

$$\Lambda = \sum_{i=1}^{k} \sum_{j=0}^{n} |\lambda^{ij}| + \sum_{l=1}^{d} |\mu^l| \le \omega_{18}(\varepsilon, N) \cdot \Lambda ,$$

where $\lim_{\varepsilon \to 0, N \to +\infty} \omega_{18}(\varepsilon, N) = 0$. Therefore, there exists $\varepsilon, N > 0$ such that $\omega_{15}(\varepsilon, N) < 1/2$, and therefore, we get that $\Lambda = 0$. As mentioned earlier, this implies that $DJ(u(z, (\delta_i)_i, (\xi_i)_i)) = 0$, and then $u(z, (\delta_i)_i, (\xi_i)_i)$ is a critical point of J for $(z, (\delta_i)_i, (\xi_i)_i) \in B_\varepsilon(0) \times \mathcal{D}_k(\varepsilon, N)$. This ends the proof of Theorem 1.1. □

References

[1] A. Ambrosetti and A. Malchiodi, *Perturbation methods and semilinear elliptic problems on* $\mathbb{R}^n$, Progress in Mathematics, vol. 240, Birkhäuser Verlag, Basel, 2006.

[2] G. Bianchi and H. Egnell, *A note on the Sobolev inequality*, J. Funct. Anal. **100** (1991), no. 1, 18–24.

[3] S. Brendle, *Blow-up phenomena for the Yamabe equation*, J. Amer. Math. Soc. **21** (2008), no. 4, 951–979.

[4] S. Brendle and F.C. Marques, *Blow-up phenomena for the Yamabe equation. II*, J. Differential Geom. **81** (2009), no. 2, 225–250.

[5] M. del Pino, M. Musso, F. Pacard, and A. Pistoia, *Large energy entire solutions for the Yamabe equation*, J. Differential Equations **251** (2011), no. 9, 2568–2597.

[6] P. Esposito, A. Pistoia, and J. Vétois, *The effect of linear perturbations on the Yamabe problem.* to appear in Math. Ann., DOI 10.1007/S00208-013-0971-9.

[7] M. Falaleev, B. Loginov, N. Sidorov, and A. Sinitsyn, *Lyapunov–Schmidt methods in nonlinear analysis and applications*, Mathematics and its Applications, vol. 550, Kluwer Academic Publishers, Dordrecht, 2002.

[8] Y. Guo, B. Li, and J. Wei, *Large energy entire solutions for the Yamabe type problem of polyharmonic operator*, J. Differential Equations **254** (2013), no. 1, 199–228.

[9] A.M. Micheletti, A. Pistoia, and J. Vétois, *Blow-up solutions for asymptotically critical elliptic equations*, Indiana Univ. Math. J. **58** (2009), no. 4, 1719–1746.

[10] A. Pistoia and J. Vétois, *Sign-changing bubble towers for asymptotically critical elliptic equations on Riemannian manifolds*, J. Differential Equations **254** (2013), no. 11, 4245–4278.

[11] O. Rey, *The role of the Green's function in a nonlinear elliptic equation involving the critical Sobolev exponent*, J. Funct. Anal. **89** (1990), no. 1, 1–52.

[12] F. Robert and J. Vétois, *Sign-Changing Blow-Up for Scalar Curvature Type Equations*, Comm. in Partial Differential Equations **38** (2013), no. 8, 1437–1465.

Frédéric Robert
Institut Élie Cartan
Université de Lorraine
BP 70239
F-54506 Vandœuvre-lès-Nancy, France
e-mail: `frederic.robert@univ-lorraine.fr`

Jérôme Vétois
Université de Nice Sophia Antipolis
Laboratoire J.A. Dieudonné
CNRS UMR 7351, Parc Valrose
F-06108 Nice Cedex 2, France
e-mail: `vetois@unice.fr`

Concentration Analysis and Applications to PDE
Trends in Mathematics, 117–141

Concentration Analysis and Cocompactness

Cyril Tintarev

Abstract. Loss of compactness that occurs in may significant PDE settings can be expressed in a well-structured form of profile decomposition for sequences. Profile decompositions are formulated in relation to a triplet (X, Y, D), where X and Y are Banach spaces, $X \hookrightarrow Y$, and D is, typically, a set of surjective isometries on both X and Y. A profile decomposition is a representation of a bounded sequence in X as a sum of elementary concentrations of the form $g_k w$, $g_k \in D$, $w \in X$, and a remainder that vanishes in Y. A necessary requirement for Y is, therefore, that any sequence in X that develops no D-concentrations has a subsequence convergent in the norm of Y. An imbedding $X \hookrightarrow Y$ with this property is called D-cocompact, a property weaker than, but related to, compactness. We survey known cocompact imbeddings and their role in profile decompositions.

Mathematics Subject Classification (2010). Primary 35J20, 35J60, 35B44, 35H20; Secondary 46B50, 58J05, 58J70, 74H35.

Keywords. Cocompact imbeddings, concentration, profile decompositions, compactness, weak convergence, wavelets.

1. Introduction

Convergence of functional sequences is easy to obtain in problems where one can invoke compactness, typically via a compact imbedding $X \hookrightarrow Y$ of two Banach spaces. In many cases, however, one deals with problems in functional spaces that possess some non-compact invariance, such as translational or scaling invariance, which produces non-compact orbits and thus makes any invariant imbedding trivially non-compact. Fortunately, the same set of invariances that destroys compactness can be employed to restore it. This approach, historically called the concentration compactness principle, emerged in the 1980's from the analysis of concentration phenomena by Uhlenbeck, Brezis, Coron, Nirenberg, Aubin and Lions, and remains widely used, in a standardized formulation, in terms of sequences of measures, due to Willem and Chabrowski. The early concentration analysis has been then refounded by two systematic theories developed in mutual isolation, a functional-analytic one ([45], with origins in [49]), and a wavelet-based one, in function spaces, (see Bahouri, Cohen and Koch [7] whose origins are in Gérard's paper [28]).

The purpose of this survey is to present current results of general concentration analysis as well as some areas of its advanced applications, such as elliptic problems of Trudinger–Moser type and "mass-critical" dispersive equations, where cocompactnes of Strichartz imbeddings was proved and employed by Terence Tao and his collaborators.

The key element of the cocompactness theory is the premise that compactness of imbeddings of two functional spaces, which in many cases is attributed to the scaling invariance $u \mapsto t^r u(t\cdot)$ (that leads to localized non-compact sequences of "blowups" or "bubbles" $t_k^r w(t_k\cdot)$, $t_k \to \infty$), can be caused by invariance with respect to any other group of operators acting isometrically on two imbedded spaces $X \hookrightarrow Y$. Furthermore, proponents of the cocompactness theory insist that there are many concrete applications which involve such operators ("gauges" or "dislocations") that are quite different from the Euclidean blowups. Indeed, recent literature contains concentration analyis of sequences in Sobolev and Strichartz spaces, involving actions of anisotropic or inhomogeneous dilations, of isometries of Riemannian manifolds (or more generally, of conformal groups on sub-Riemannian manifolds and other metric structures) and of transformations in the Fourier domain.

Improvement of convergence, based on elimination of concentration (understood in the abstract sense as terms of the form $g_k w$, where $\{g_k\}$ is a non-compact sequence of gauges) can be illustrated in the sequence spaces on an elementary example based on Proposition 1 of [29].

Example. The imbedding $\ell^p(\mathbb{Z}) \hookrightarrow \ell^q(\mathbb{Z})$, $1 \le p < q \le \infty$, is not compact, since sequences of the form $u(\cdot + k)$ converge to zero weakly in ℓ^p, but have constant ℓ^∞-norm. Let us decide that $u(\cdot + j_k)$ with $|j_k| \to \infty$ is a typical "concentrating behavior" and eliminate it from a given sequence $u_k \in \ell^p$ by assuming that for any sequence $j_k \in \mathbb{Z}$, one has $u_k(\cdot + j_k) \rightharpoonup 0$ in ℓ^p. Then $u_k(j_k) \to 0$ in $\mathbb{R}$ for any sequence $j_k \in Z$, which implies $u_k \to 0$ in ℓ^∞. Since $\|u\|_q^q \le \|u\|_\infty^{q-p}\|u\|_p^p$, one also has $u_k \to 0$ in ℓ^q for any $q > p$. We conclude that elimination of possible "concentration" caused by shifts $u \mapsto u(\cdot + j)$, $j \in \mathbb{Z}$, assures convergence in ℓ^q.

This example gives motivation to the following definitions.

Definition 1.1. (Gauged weak convergence.) Let X be a Banach space, and let D be a bounded set of bounded linear operators on X containing the identity operator. One says that a sequence $u_k \in D$ converges to zero D-weakly if $g_k u_k \rightharpoonup 0$ with any choice of sequence $(g_k) \subset D$. We denote the gauged weak convergence as $u_k \overset{D}{\rightharpoonup} 0$.

Definition 1.2. Let X be a Banach space continuously imbedded into a Banach space Y. One says that the imbedding $X \hookrightarrow Y$ is cocompact (relative to the set D) if $u_k \overset{D}{\rightharpoonup} 0$ implies $\|u_k\|_Y \to 0$.

Definition 1.3. One says that the norm of Y provides a (local) metrization of the D-weak convergence on X, if for any bounded sequence $u_k \in X$,

$$u_k \overset{D}{\rightharpoonup} 0 \Longleftrightarrow \|u_k\|_Y \to 0.$$

Note that we speak only about *local* metrization, on the balls of X, and that convergence in different norms, for sequences restricted to such balls, becomes equivalent. For example, all ℓ^q-convergences with $q > p \geq 1$ are equivalent on a ball of ℓ^p, and all L^q-convergences with $q \in (p, \frac{pN}{N-p})$ are equivalent on a ball of $W^{1,p}(\mathbb{R}^N)$, $1 \leq p < N$ (by the limiting Sobolev imbedding and the Hölder inequality).

Example. Let $X = H^1_{0,\mathrm{rad}}(B)$ be the subspace of all radial functions in the Sobolev space $H^1_0(B)$, let $Y = L^6(B)$, where $B \subset \mathbb{R}^3$ is a unit ball, and let $D = \{h_t u(x) = t^{1/2}u(tx)\}_{t>0}$.
The lack of compactness of the imbedding $X \hookrightarrow Y$ is demonstrated by the blowup sequences $t_k^{1/2}w(t_k x)$ with $t_k \to \infty$ and $w \in \mathcal{D}^{1,2}_{rad}(\mathbb{R}^3) \setminus \{0\}$. If, however, u_k is a bounded sequence in X and all its "deflation sequences" $t_k^{-1/2}u_k(t_k^{-1}x)$, with $t_k \to \infty$, converge to zero weakly in X, the sequence u_k has a subsequence convergent in Y. This we express as D-cocompactness of the imbedding $H^1_{0,\mathrm{rad}}(B) \hookrightarrow L^6(B)$. See Proposition 2.3 below with an elementary proof.

While the term *cocompact imbedding* is recent, the property itself has been known for Sobolev spaces for decades, and can traced to a lemma by Lieb [35]. Cocompactness of imbeddings relative to rescalings (actions of translations and dilations), which for the Sobolev spaces is usually credited to Lions [38], is known today for a range of Besov and Triebel–Lizorkin spaces as X and different Besov, Triebel–Lizorkin, Lorentz and BMO spaces as Y ([7], once we note that the crucial Assumption 1 in this paper, expressed in terms of the wavelet bases implies cocompactness and is in a general case equivalent to it, see the argument in Subsection 2.2 below). Cocompactness of Sobolev imbeddings is known also for function spaces on manifolds, with the role of translations and dilations taken over by the conformal group. Cocompactness is also established in Trudinger–Moser imbeddings (with inhomogeneous dilations) and in Strichartz imbeddings in dispersive equations. Section 2 gives a summary of known cocompact imbeddings.

Section 3 studies profile decompositions that arise in the presence of cocompact imbeddings. These are largely functional-analytic results, established in two general cases, for Hilbert spaces and for Banach spaces imbedded into L^p-spaces. Beyond that, profile decompositions have been proved for a wide range of imbeddings for spaces that have a wavelet basis of rescalings, presented in [7] (with a remainder that vanishes in a weaker sense than in the Y-norm), as well as for some Trudinger–Moser and Strichartz imbeddings. First profile decompositions in literature were found for specific sequences, typically, Palais–Smale sequences for semilinear elliptic functionals (Struwe [52], Brezis and Coron [16], Lions himself [40] and Benci & Cerami [12]). The first profile decomposition for *general* bounded sequences in $\mathcal{D}^{1,p}(\mathbb{R}^N)$ was proved by Solimini [49], and, furthermore, from Solimini's argument it also became clear that a profile decomposition is essentially a functional-analytic phenomenon, and it is the cocompactness (still not a named property) that requires a substantial hard analysis.

Section 4 deals with reduction of cocompactness and profile decomposition to subspaces, which in many cases results in more specific types of concentration or even its disappearance, as, for example, in case of the radial subspaces of Sobolev spaces (Strauss Lemma), as well as provides few sample arguments that allow to derive cocompactness by transitivity of imbeddings or interpolation.

The range of applications of concentration analysis is wider than the scope of this survey. We do not consider here in any great detail time evolution of concentration in the initial data that arises in semilinear dispersive equations, large-time emergence of concentration in evolution equations, applications to geometric problems, and blow-up arguments for sequences of solutions of PDE that may benefit from further built-in structure of the equations.

2. Cocompact imbeddings

2.1. Example: cocompactness of Sobolev imbeddings with elementary poofs

The following theorem is essentially a lemma of Lieb from [35].

Theorem 2.1. *Let*

$$D = \{u \mapsto u(\cdot - y),\ y \in \mathbb{Z}^N\}. \tag{2.1}$$

Assume that $N > p > 1$. *The Sobolev imbedding* $W^{1,p}(\mathbb{R}^N) \hookrightarrow L^q$, $p \le q \le p^* = \frac{pN}{N-p}$, *is cocompact relative to the group* D, *unless* $q = p$ *or* $q = p^*$. *The* L^q*-norm gives a* $W^{1,p}$*-local metrization of the* D*-weak convergence.*

The original statement asserted only convergence in measure, but under a $W^{1,p}$-bound, so that convergence in measure implies convergence in L^q, $p \le q \le p^* = \frac{pN}{N-p}$. Moreover, the original statement (and many other cocompactness statements in literature) is expressed in terms of negation: if a bounded sequence in $W^{1,p}$ does not converge in measure, then it has a subsequence with a nonzero, under suitable translations, weak limit.

Proof. Let $Q = (0,1)^N$. Assume that and $u_n(\cdot - y_n) \rightharpoonup 0$ in $W^{1,p}(\mathbb{R}^N)$ for any sequence $y_n \in \mathbb{Z}^N$. For every $y \in \mathbb{Z}^N$we have by the Sobolev inequality

$$\int_{Q+y} |u_n|^q \le C \int_{Q+y} (|\nabla u_n|^p + |u_n|^p) \left(\int_{Q+y} |u_n|^q\right)^{1-p/q}.$$

Adding up the inequalities over $y \in \mathbb{Z}^N$ while estimating the last factor by the supremum over y, and replacing the supremum by twice the value of the term at the "worst" values of y_n, we get

$$\int_{\mathbb{R}^N} |u_n|^q \le C \left(\int_Q |u_n(\cdot - y_n)|^q\right)^{1-p/q},$$

which converges to zero as a consequence of compactness of the subcritical Sobolev imbedding over Q. □

This is a model proof of cocompactness that uses partitioning of the norm of the target space into "cells of compactness", in this case involving the fundamental domain of the lattice group acting on $\mathbb{R}^N$, where one can benefit from compactness, followed by a reassembly of the full norm from the vanishing terms. In the wavelet approach presented in Subsection 2.2 the role similar to that of "compactness cell" is played by the mother wavelet.

The limiting Sobolev imbedding $\mathcal{D}^{1,p}(\mathbb{R}^N) \hookrightarrow L^{\frac{pN}{N-p}}(\mathbb{R}^N)$ is cocompact only if one enlarges the group D by the action of dilations. The following statement originates in Lions [38] with four different proofs given later by Solimini [49]; Gérard [28] (for $p = 2$) and Jaffard [29]; the author [56]; and Kilip and Visan [32].

Theorem 2.2. *Let $\gamma > 0$, $N > p \geq 1$, $r = \frac{N-p}{p}$, and let*

$$D = \{u \mapsto \gamma^{rj} u(\gamma^j(\cdot - y)),\ y \in \mathbb{R}^N, j \in \mathbb{Z}\}. \tag{2.2}$$

The imbedding $\mathcal{D}^{1,p}(\mathbb{R}^N) \hookrightarrow L^{\frac{pN}{N-p}}$ is cocompact relative to the group D. Furthermore, the $L^{\frac{pN}{N-p}}$-norm on bounded subsets of $\mathcal{D}^{s,p}(\mathbb{R}^N)$ provides a metrization of D-weak convergence.

This result is usually rendered in literature with the scaling factor in $(0, \infty)$ instead of the discrete subset $\gamma^{\mathbb{Z}}$. Sufficiency of the discrete values can be observed by following the available proofs, or derived a posteriori. We give below a proof of the statement reduced to the case of the radial subspace of $\mathcal{D}^{1,p}$.

Proposition 2.3. *Let $\gamma > 1$, $N > p > 1$, and let*

$$D = \{u \mapsto \gamma^j u(\gamma^{\frac{p}{N-p}j}\ \cdot),\ j \in \mathbb{Z}\}.$$

The limiting Sobolev imbedding of the radial subspace of $\mathcal{D}^{1,p}(\mathbb{R}^N)$, $\mathcal{D}^{1,p}_{\mathrm{rad}}(\mathbb{R}^N) \hookrightarrow L^{\frac{pN}{N-p}}(\mathbb{R}^N)$, is cocompact relative to the group D and the $L^{\frac{pN}{N-p}}$-norm provides a $\mathcal{D}^{1,p}(\mathbb{R}^N)$-local metrization of the D-weak convergence.

Proof. Let $u_k \in \mathcal{D}^{1,p}_{\mathrm{rad}}(\mathbb{R}^N)$ and assume that with any $j_k \in \mathbb{Z}$, $\gamma^{-\frac{N-p}{p}j_k} u_k(\gamma^{j_k}\cdot) \rightharpoonup 0$. Let $p^* = \frac{pN}{N-p}$ and apply the limiting Sobolev inequality to $\chi_j(u) = \gamma^j \chi(\gamma^{-j}|u|)$, where $\chi \in C_0^\infty((\frac{1}{\gamma}, \gamma^2))$ satisfying $\chi(s) = s$ for $s \in [1, \gamma]$:

$$\left(\int_{|u_k| \in (\gamma^j, \gamma^{j+1})} |u_k|^{p^*} \mathrm{d}x\right)^{\frac{p}{p^*}} \leq C \int_{|u_k| \in (\gamma^{j-1}, \gamma^{j+2})} \left(|\nabla u_k|^p + \frac{|u_k|^p}{|x|^p}\right) \mathrm{d}x.$$

Taking the sum over $j \in \mathbb{Z}$ and estimating the last integrand in the right-hand side by the Hardy inequality, we have

$$\int_{\mathbb{R}^N} |u_k|^{p^*} \leq C \int_{\mathbb{R}^N} |\nabla u_k|^p \left(\sup_{j \in \mathbb{Z}} \int_{|\gamma^j u_k(\gamma^{\frac{p}{N-p}j} x)| \in (1,\gamma)} |\gamma^j u_k(\gamma^{\frac{p}{N-p}j} x)|^{p^*} \mathrm{d}x\right)^{1 - \frac{p}{p^*}}.$$

Replace the supremum in the right-hand side by twice the value at a suitable sequence j_k. The integrand in this term is a radial function, uniformly bounded in $L^1 \cap L^\infty$, which, as a rescaled weakly vanishing sequence, converges to zero pointwise. Consequently, the left-hand side converges to zero and cocompactness is proved.

To verify the local metrization, assume that $u_k \to 0$ in $L^{p^*}(\mathbb{R}^N)$ and that the sequence is bounded in $\mathcal{D}^{1,p}$. Then, for any sequence $g_k \in D$, the scaling invariance of the L^{p^*}-norm implies that $g_k u_k \rightharpoonup 0$ in L^{p^*}. Then, by density, $g_k u_k \rightharpoonup 0$ in $\mathcal{D}^{1,p}$, and thus $u_k \overset{D}{\rightharpoonup} 0$. □

2.2. Cocompactness relative to rescalings: the wavelet approach

Imbeddings of Sobolev type are cocompact relative to the rescalings group (2.2) for a large class of spaces. The first results of this type were obtained for imbeddings of Riesz potential spaces $\dot{H}^{s,p}(\mathbb{R}^N)$ into $L^{\frac{pN}{N-sp}}(\mathbb{R}^N)$, $s > 0$, $1 < p < N$, Gérard [28] (for Sobolev spaces with $p = 2$) and Jaffard [29] (for general p). Here we summarize a major generalization of their analysis in Bahouri, Cohen and Koch [7].

Let $X \hookrightarrow Y$ be two Banach spaces of functions on $\mathbb{R}^N$, and assume that there exists an unconditional Schauder wavelet basis of wavelets $\Gamma\psi$, same for X and Y, where

$$\Gamma = \{u \mapsto g_{j,y}u = 2^{rj}u(2^j \cdot -y)\}_{j\in\mathbb{Z},y\in\mathbb{Z}^N}, \tag{2.3}$$

and $r > 0$ and $\psi \in X$ ("mother wavelet") are fixed. It's assumed that operators $u \mapsto u(\cdot - y)$, $y \in \mathbb{R}^N$, and $u \mapsto t^r u(t\cdot)$, $t > 0$, are isometries in both X and Y. For each $M \in \mathbb{N}$ and for every function $u \in X$, expanded in the basis $\Gamma\psi$ as $u = \sum_{g\in\Gamma} c(g)g\psi$, define a subset $\Gamma_M(u) \subset \Gamma$ of cardinality M which corresponds to the M largest values of $|c(g)|$, and set

$$Q_M u = \sum_{g\in\Gamma_M(u)} c(g)g\psi.$$

Note that such set Γ_M always exists (and is not unique when some $|c(g)|$ are equal, so one fixes it arbitrarily) due to the fact that $\Gamma\psi$ is a Schauder basis for X, and thus for any $\eta > 0$ only finitely many coefficients $c(g)$ have their absolute value larger than η.

The main condition in [7], required for the imbedding $X \hookrightarrow Y$ (Assumption 1), is

$$\sup_{\|u\|_X\le 1} \|Q_M u - u\|_Y \to 0 \text{ as } M \to \infty. \tag{2.4}$$

Remark 2.4. When X is reflexive, condition (2.4) implies that the imbedding $X \hookrightarrow Y$ is D-cocompact, where D is the rescaling group (2.2) with $\gamma = 2$. Indeed, assume that $h_k u_k \rightharpoonup 0$ in X for every rescaling sequence $\{h_k\} \subset D$. Let $\Gamma' = \{g^{-1} : g \in \Gamma\} \subset D$, and note that $\Gamma'\Gamma\psi \subset \Gamma\psi$. Consider the wavelet expansion coefficients $c_k(g)$, $g \in \Gamma$, of functions u_k, and let $g_k \in \Gamma$ be such that $|c_k(g_k)| = \eta_k := \max_{g\in\Gamma} |c_k(g)| = \max_{g\in\Gamma_M} |c_k(g)|$. Then the coefficient $c_k'(\mathrm{id})$, corresponding to the basis vector ψ in the wavelet expansion for $g_k^{-1}u_k$, is equal to $c(g_k)$. At the same time $c'_k(\mathrm{id}) \to 0$ since the coefficients of expansions in a Schauder basis of a reflexive space are continuous linear functionals and $g_k^{-1}u_k \rightharpoonup 0$. Fix now $\epsilon > 0$ and let $M = M(\epsilon)$ be such that $\|Q_M u_k - u_k\|_Y \leq \epsilon$. Then

$$\|u_k\|_Y = \|Q_M u_k + (u_k - Q_M u_k)\|_Y \leq \|Q_M u_k\|_Y + \epsilon \leq M(\epsilon)\eta_k + \epsilon.$$

Let $k \to \infty$ and note that ϵ was arbitrary.

Conversely, cocompactness of the imbedding relative to rescalings easily implies (2.4) if the space X and its basis $\Gamma\psi$ satisfy the following condition: $Q_M u_M - u_M \rightharpoonup 0$ as $M \to \infty$ for any bounded sequence $\{u_M\} \subset X$. Details are left to the reader.

For spaces brought up below, existence of an unconditional Schauder basis of rescaling wavelets is known, and the norms in their spaces have equivalent definitions in terms of expansion coefficients in the wavelet basis (see [41]). Using these characterizations, [7] verifies condition (2.4), and thus cocompactness of the imbeddings whenever X is reflexive. Results of [7] contain several earlier wavelet-based cocompactness results, in particular, [28, 29, 23, 33]. Some cases follow from the others via a simple transitivity argument: if $X \hookrightarrow Y$, $Y \hookrightarrow Z$ and one of the imbeddings satisfies 2.4 (or is cocompact), then satisfies 2.4 (resp. is cocompact), or from the equivalence of different Y-convergences on bounded sets of X.

Theorem 2.5. *Let $p, q \in [1, \infty]$, $a, b \in (0, \infty]$, $s > t \geq 0$, and $r = \frac{N}{p} - s > 0$. The following imbeddings satisfy (2.4) (and are cocompact relative to rescalings whenever the domain is a reflexive space).*

(i) $\dot{B}^s_{p,p}(\mathbb{R}^N) \hookrightarrow L^q(\mathbb{R}^N)$, $q < \infty$, $\frac{1}{p} - \frac{1}{q} = \frac{s}{N} > 0$.
(ii) $\dot{B}^s_{p,p}(\mathbb{R}^N) \hookrightarrow \dot{B}^t_{q,q}(\mathbb{R}^N)$, $\frac{1}{p} - \frac{1}{q} = \frac{s-t}{N} > 0$.
(iii) $\dot{B}^s_{p,q}(\mathbb{R}^N) \hookrightarrow \dot{F}^t_{q,b}(\mathbb{R}^N)$, $\frac{1}{p} - \frac{1}{q} = \frac{s-t}{N} > 0$.
(iv) $\dot{B}^s_{p,a}(\mathbb{R}^N) \hookrightarrow \dot{B}^t_{q,b}(\mathbb{R}^N)$, $\frac{1}{p} - \frac{1}{q} = \frac{s-t}{N} > 0$, $a < b$ *(no cocompactness when $a = b$).*
(v) $\dot{B}^s_{p,p}(\mathbb{R}^N) \hookrightarrow \mathrm{BMO}(\mathbb{R}^N)$, $s = \frac{N}{p} > 0$.
(vi) $\dot{B}^s_{p,a}(\mathbb{R}^N) \hookrightarrow L^{q,b}(\mathbb{R}^N)$, $\frac{1}{p} - \frac{1}{q} = \frac{s}{N} > 0$, $a < b$.
(vii) $\dot{F}^s_{p,a}(\mathbb{R}^N) \hookrightarrow \dot{F}^t_{q,b}(\mathbb{R}^N)$, $\frac{1}{p} - \frac{1}{q} = \frac{s-t}{N} > 0$, $a, b > 0$.

Note that for $m \in \mathbb{N}$, the space $\dot{F}^m_{p,2}$ coincides with the Sobolev space $\mathcal{D}^{m,p}(\mathbb{R}^N)$, and $\dot{F}^0_{q,2}$ coincides with L^q, so the last imbedding includes the limiting Sobolev imbeddings. Profile decompositions for these imbeddings, which we discuss in Section 3, are also given in [7].

2.3. Cocompactness of trace imbeddings

Theorem 2.6. *Let $1 < p < N$ and $\bar{p} = \frac{p(N-1)}{N-p}$. The following imbeddings are cocompact:*

(i) $W^{1,p}(\mathbb{R}^N) \hookrightarrow L^q(\mathbb{R}^{N-1})$, $q \in (p, \bar{p})$, *relative to the lattice shifts* $\{u \mapsto u(\cdot - y)\}_{y \in \mathbb{R}^{N-1}}$.
(ii) $\mathcal{D}^{1,p}(\mathbb{R}^N) \hookrightarrow L^{\bar{p}}(\mathbb{R}^{N-1})$ *relative to the rescalings group* (2.2).

Proof. The proof of (i) is repetitive of the argument in the paragraph 2.3.1 above, employing the cubic lattice neighborhood of the hyperplane, instead of the cubic tessellation of the whole $\mathbb{R}^N$. The proof of (ii) for $p = 2$ is given in Lemma 5.10 in [56], and extends trivially to general $p > 1$. □

Results in both cases easily extend to the case of hyperplanes of any dimension $d > p$. In both cases the corresponding L^p-norms give metrization of D-weak convergence. We are not aware of further results in literature on cocompactness of trace imbeddings, but it is plausible that most trace imbeddings for Besov and Triebel–Lizorkin spaces are similarly cocompact, and the wavelet argument of [7] can be applied here as well.

2.4. Cocompactness of Sobolev imbeddings on metric structures

For the subcritical Sobolev imbeddings on manifolds, the actions of isometries play the same role in cocompactness as parallel translations in the Euclidean case. Let M be a complete Riemannian manifold and let $I(M)$ be its isometry group. The following result deals with cocompactness of "magnetic" Sobolev spaces, corresponding in appropriate cases to the Schrödinger operator with external magnetic field. Magnetic Sobolev space $W^{1,p}_\alpha(M)$, with a fixed $\alpha \in T^*M$, called magnetic potential, is the space of functions $M \to \mathbb{C}$ with measurable weak derivatives, characterized by the finite norm

$$\|u\|^p_{1,p} = \int_M (|du + iu\alpha|^p + |u|^p).$$

The usual Sobolev space corresponds to the form α being exact (i.e., zero modulo gauge transformation). By the diamagnetic inequality $|du + i\alpha u| \geq |d|u||$, so one always has $W^{1,p}_\alpha(M) \hookrightarrow W^{1,p}(M)$. When α is not exact, assume in addition that M is simply connected. Then for each diffeomorphism $\eta : M \to M$, such that $d(\alpha \circ \eta) = d\alpha$, there exists a unique real-valued function $\varphi_\eta \in C^\infty(M)$ satisfying $d\varphi = \alpha \circ \eta - \alpha$. An elementary computation shows that if $\eta \in I(M)$, then the $W^{1,p}_\alpha(M)$-norm is preserved by the operators $u \mapsto e^{i\varphi_\eta} u \circ \eta$, called *magnetic shifts* (see [6]). In physics, the meaning of the relation $d\alpha \circ \eta = d\alpha$ is that the magnetic field $d\alpha$ is periodic under isometries of M.

Theorem 2.7. *Let M be a complete smooth non-compact Riemannian N-manifold, cocompact (periodic) relative to $I(M)$, and simply connected whenever $\alpha \in T^*M$ is not exact. Let G be a closed subgroup of $I(M)$ and let $Q \subset M$ be a bounded set such that $GQ = M$. Define*

$$D_G = \{u \mapsto e^{i\varphi_\eta} u \circ \eta,\ \eta \in G\}. \tag{2.5}$$

Then the imbedding $W^{1,p}_\alpha(M) \hookrightarrow L^q(M)$, $1 < p \leq q \leq p^* = \frac{pN}{N-p}$, $N > p$, is cocompact relative to the set of operators D_G. Moreover, the L^q-norm provides metrization of the D-weak convergence on bounded subsets of $W^{1,p}_\alpha$.

Proof. The proof of cocompactness and verification of metrization for $p = 2$ is given in [56], Lemma 9.4. The proof for general p is completely analogous. The argument is similar to that for Theorem 2.1, and involves existence of a covering for M of uniformly finite multiplicity by sets $\{\eta V\}_{\eta\in G'}$, with some set $G' \subset G$ and some open set $V \subset M$. We give here an argument for reduction of the general (magnetic) case to the case $\alpha = 0$.

If $e^{i\varphi_{\eta_k}} u_k \circ \eta_k \rightharpoonup 0$ in $W^{1,p}_\alpha(M)$, then $|u_k \circ \eta_k|$ converges to zero a.e. and, by the diamagnetic inequality, is bounded in $W^{1,p}(M)$. Then, from cocompactness of the "non-magnetic" imbedding $W^{1,p}(M) \hookrightarrow L^q(M)$, follows $|u_k| \to 0$ in $L^q(M)$, $q \in (p,p^*)$. □

Cocompactness of subcritical imbeddings relative to isometries extends to Sobolev spaces on metric structures other than Riemannian manifolds, in particular, to the Sobolev spaces of the Kohn Laplacian on Carnot groups ([50]) and of blowups of a self-similar fractal of a class involving Sierpinski gasket, as in [55]; see also the paper [14] in the setting of axiomatic Sobolev spaces on general metric structures.

Let now M be a simply connected nilpotent stratified Lie group (Carnot group) with a stratification $T_eM = \bigoplus_{j=1}^m Y_j$. Let $\nu = \sum_{j=1}^m j \dim Y_j \geq 3$. One calls the diffeomorphism $T_s = \exp_e \tau_s \exp_e^{-1}$ of M an *anisotropic dilation* if τ_s is given by $\tau_s|_{Y_j} = s^j$, $j = 1,\dots,m$. Let $v_1, \dots v_{\dim Y_1} \in T_eM$ be an orthonormal basis in Y_1 and consider the subelliptic Sobolev space $\dot{H}^1(M)$, characterized by the norm

$$\|u\|^2 = \int_M \left| \sum_{i=1}^{\dim Y_1} |\langle du, v_i\rangle|^2 \right|.$$

In particular, for the Heisenberg group $\mathbb{H}_N$, identified as $\mathbb{R}^N \times \mathbb{R}^N \times \mathbb{R}$ with the group law

$$(x,y,t)(x',y',t') = (x+x', y+y', t+t'+2x\cdot y' - 2x'\cdot y),$$

the first stratum is $Y_1 = \mathbb{R}^N \times \mathbb{R}^N$, the effective dimension ν equals $2N+2$, the anisotropic dilations are $T_s(x,y,t) = (sx, sy, s^2t)$, and the Sobolev norm is given by

$$\|u\|^2 = \int_{\mathbb{R}^{2N+1}} \left(\sum_{i=1}^N |(\partial_{x_i} + 2y_i\partial_t)u|^2 + \sum_{i=1}^N |(\partial_{y_i} - 2x_i\partial_t)u|^2 \right) dxdydt.$$

The space $\dot{H}^1(M) \hookrightarrow L^{\frac{2\nu}{\nu-2}}(M)$ is continuously imbedded into $H^1(M)$. For further details on subelliptic Sobolev spaces we refer to ([25], [26], as well a brief exposition in [56], Chapter 9.

Theorem 2.8. *Let $\dot{H}^1(M)$ be the subelliptic Sobolev space on the Carnot group M as above, let $\gamma > 1$ and let*

$$D = \{u \mapsto \gamma^{\frac{\nu-2}{2}j} u \circ \eta \circ T_{\gamma^j},\ \eta \in M,\ j \in \mathbb{Z}\}.$$

Then the imbedding $\dot{H}^1(M) \hookrightarrow L^{\frac{2\nu}{\nu-2}}(M)$ is cocompact relative to D. Furthermore, the $L^{\frac{2\nu}{\nu-2}}$-norm provides a metrization of D-weak convergence on bounded subsets of $\mathcal{D}^{1,p}(\mathbb{R}^N)$.

This result is proved in [56], Lemma 9.14, for $p = 2$ (extension to $p \neq 2$ is elementary). Earlier results can be found in [50] for the subcritical case and the general Carnot group, and in [11] for the case of Heisenberg group). Like in the Euclidean case, many applications to subelliptic PDE on Lie groups could be handled with the help of the Willem–Chabrowski version of concentration compactness (e.g., in papers of Garofalo et al) or with Struwe's "global compactness" ([52], see Theorem 3.1 in the book [24] of Hebey, Druet and Robert), which is a realization of a possible more general profile decomposition for manifolds, for the case of Palais–Smale sequences with dilations expressed via the exponential map.

2.5. Cocompactness of the Moser–Trudinger imbedding

The counterpart of Sobolev imbedding of $\mathcal{D}^{1,p}(\mathbb{R}^N)$ in the borderline case $p = N$, is the imbedding defined by the Moser–Trudinger inequality (Yudovich, [58], independently rediscovered by Pohozhaev, Peetre and Trudinger, and with the optimal exponent proved by Moser [42]). Moser–Trudinger inequality is stated for bounded domains and it is false for $\mathbb{R}^N$. There reason for that is that the gradient norm $\|\nabla u\|_N$ on $C_0(\mathbb{R}^N)$ does not dominate any linear functional $\langle \varphi, u\rangle$ with $\varphi \in \mathcal{D}'(\mathbb{R}^N) \setminus \{0\}$, i.e., the space $\mathcal{D}^{1,N}(\mathbb{R}^N)$ defined as a formal completion of $C_0^\infty(\mathbb{R}^N)$ in the gradient norm, is not continuously imbedded even into the space of distributions. Another way to express this is that there exists a Cauchy sequence representing zero of the completion, which converges to 1 uniformly on every compact set. A counterpart of the Moser–Trudinger inequality for $\mathbb{R}^N$, proved by Li and Ruf [34], involves the full Sobolev norm rather than the gradient norm, and it also avoids lower powers of u which have poor integrability at infinity. In what follows we will use the Sobolev norm $\|u\|_{1,N}$ equal to the standard Sobolev norm if the domain $\Omega \subset \mathbb{R}^N$ is unbounded and to the equivalent norm of $W_0^{1,N}(\Omega)$, $\|\nabla u\|_N$, if Ω is bounded. The inequality that expresses both the Moser–Trudinger inequality and the Li–Ruf inequality, is

$$\sup_{u\in W_0^{1,N}(\Omega),\|u\|_{1,N}\leq 1} \int_\Omega \exp_N(\alpha_N |u|^{N'})dx < \infty,$$

where $\alpha_N = N\omega_{N-1}^{1/(n-1)}$, ω_{N-1} is the measure of the unit sphere in $\mathbb{R}^N$, $N' = \frac{N}{N-1}$ and $\exp_N(t) = e^t - \sum_{j=0}^{N-2} \frac{t^j}{j!}$. For bounded domains one may equivalently use e^t instead of $\exp_N(t)$, since the subtracted polynomial also has a bounded integral.

The imbedding expressed by the Moser–Trudinger inequality is $W_0^{1,N}(\Omega) \hookrightarrow \exp_N L^{N'}$, where $\exp_N L^{N'}$ is the Orlicz space space associated with the convex function $\exp_N(t^{N'})$.

In the spaces of radial functions cocompactness of imbeddings is established in the cases when Ω is a disk (without loss of generality we consider here the unit disk B) or $\mathbb{R}^N$ (if Ω is an annulus or an exterior disk, the imbedding is compact for elementary reasons). The result below is due to [2]:

Theorem 2.9. *The imbedding $W^{1,N}_{0,\mathrm{rad}}(B) \hookrightarrow \exp L^{N'}(B)$ is cocompact relative to the group*

$$D = \{u \mapsto s^{1-1/N} u(r^s)\}_{s>0}.$$

Furthermore, a local metrization of the D-weak convergence is provided by the norm $\sup_{0<r<1} \frac{|u(r)|}{(\log \frac{1}{r})^{1/N'}}$.

One can easily derive from here cocompactness in $W^{1,N}_{\mathrm{rad}}(\mathbb{R}^N)$ by considering a sequence $u_k - u_k(1)$ on B, bounded in $H^1_0(B)$, and noting that the restriction of a radial sequence $u_k \rightharpoonup 0$ in $W^{1,N}(\mathbb{R}^N)$ to the complement of B has a uniform bound and converges to zero in $L^{NN'}$.

Corollary 2.10. *The imbedding $W^{1,N}_{\mathrm{rad}}(\mathbb{R}^N) \hookrightarrow \exp_N L^{N'}(\mathbb{R}^N)$ is cocompact relative to the group*

$$D = \{g_s : \; g_s u(r) = u(1) + s^{-1/N'}(u(r^s) - u(1)), \; r \le 1; \; g_s u(r) = u(r), \; r > 1\}_{s>0},$$

and metrization of the D- weak convergence is acheived by the norm

$$\sup_{0<r<1} \frac{|u(r) - u(1)|}{(\log \frac{1}{r})^{1/N'}} + \|u\|_{p,\mathbb{R}^N \setminus B}, p \in (N, \infty).$$

Without the assumption of radiality, a cocompactness result is known presently only for $N = 2$.

Theorem 2.11 ([4]). *The imbedding $H^1_0(B) \hookrightarrow \exp L^2(B)$ is cocompact relative to the compositions of translations and transformations*

$$u(z) \mapsto j^{-1/2} u(z^j), \; j \in \mathbb{N}, \tag{2.6}$$

where z expresses coordinates of B as a complex variable. A (quasi-)metrization of the D-weak convergence is acheived by the quasinorm $\sup_{0<r<1} \frac{u^\star(r)}{(\log \frac{1}{r})^{1/2}}$, *where $u^\star$ denotes the symmetric decreasing rearrangement of u.*

Note that the quasinorm $\sup_{0<r<1} \frac{u^\star(r)}{(\log \frac{1}{r})^{1/2}}$ is marginally stronger than the Zygmund quasinorm $\sup_{0<r<1} \frac{u^\star(r)}{(1+\log \frac{1}{r})^{1/2}}$, which is equivalent to the $\exp L^2$-norm (see [13]).

2.6. Cocompactness of Strichartz imbeddings

Consider a Strichartz imbedding that estimates the space-time L^q- norm of the solution of the evolutionary Schrödinger equation by the L^2-norm of the initial data (for details see [17, 32]). The following result is due to Terence Tao, [53], and the version of the proof in [32] optimizes the group. From the presentation of [32] we could easily infer that cocompactness remains valid if one restricts dilations to a discrete group. In the theorem below $\hat{u}$ denotes the Fourier transform.

Theorem 2.12. *Let $N \geq 3$. The imbedding defined by the inequality*

$$\|e^{it\Delta}u\|_{L^q(\mathbb{R}^{N+1})} \leq C\|u\|_{L^2(\mathbb{R}^N)},\ q = \frac{2N+2}{N},$$

is cocompact in L^2 with respect to the product group of the following transformations ($\gamma > 1$ is a fixed number):

$$u(x) \mapsto \gamma^{\frac{Nj}{2}} u(\gamma^j x),\ j \in \mathbb{Z};$$
$$u(x) \mapsto u(x-y), \quad y \in \mathbb{R}^N;$$
$$\hat{u}(\xi) \mapsto \hat{u}(\xi - \xi'), \quad \xi' \in \mathbb{R}^N.$$

Translations in the Fourier domain allows to consider functions with support on a cube in the Fourier domain, for which convergence in L^2 implies convergence in $C^m(\mathbb{R}^N)$ for any $m \in \mathbb{N}$. The main technical point in the proof is the "reassembly" of the inequality from the "cells of compactness" using methods of harmonic analysis.

3. Profile decompositions

Given an imbedding of a Banach space X into a Banach space Y and a bounded set D of bounded linear bijections $X \to X$, a profile decomposition is a representation of a bounded sequence $\{u_k\} \subset X$ in the form

$$u_k = \sum_{n \in \mathbb{N}} g_k^{(n)} w^{(n)} + r_k,\ r_k \overset{Y}{\to} 0, \tag{3.1}$$

where the terms $g_k^{(n)} w^{(n)}$, $n \in \mathbb{N}$ (which may be called *elementary concentrations*), are actions of sequences of operators $\{g_k^{(n)}\}_k$ on elements $w^{(n)} \in X$, (which may be called *concentration profiles*), that satisfy ${g_k^{(n)}}^{-1} u_k \rightharpoonup w^{(n)}$. Furthermore, the elementary concentrations are expected to be asymptotically decoupled in the sense that ${g_k^{(n)}}^{-1} g_k^{(m)} \rightharpoonup 0$, $m \neq n$. Convergence of the remainder in X should not be generally expected, as can be illustrated on the following example. Let $X = \ell^p(\mathbb{Z})$ and let $u_k(j) = 1/k^{1/p}$ for $j = 1, \dots, k$ taking zero values for all other j. If D is the set of shifts $u \mapsto u(\cdot - j)$, $j \in \mathbb{Z}$, then u_k has no (nonzero) profiles under any shift sequence, so $r_k = u_k$ and $\|r_k\|_p = 1$. At the same time $\|r_k\|_q \to 0$ for any $q > p$. Many profile decompositions found in literature (typically in papers using the wavelet argument, starting with [28]) are stated with a weaker remainder, namely, in the form

$$u_k = \sum_{n=1}^{M} g_k^{(n)} w^{(n)} + r_k^{(M)},\ \lim_{M \to \infty} \limsup_{k \to \infty} \|r_k^{(M)}\|_Y = 0. \tag{3.2}$$

In many significant cases, however, such as Palais–Smale sequences for elliptic problems, one can establish a lower bound on some norm of the concentration profiles, which assures that only finitely many profiles are non-zero, in which case the weak remainder $r_k^{(M)}$ with M equal to the number of nonzero profiles, is the same as the strong remainder.

As it was established in [45], profile decompositions hold whenever X is a Hilbert space under a general condition on the set D. A similar result for Banach space is also expected to be true, under some, rather weak but yet unknown general conditions on X. As a temporary fix we give here a profile decomposition in the case when X is continuously imbedded into $L^q(M)$, where M is a measure space.

The abstract profile decomposition gives a remainder r_k that converges to zero D-weakly. Cocompactness is defined as the property of X, Y, D that D-weak convergence in X implies convergence in the norm of Y. It is verified (although not under that name and often expressed in different terms) and employed also in the proofs of profile decompositions for specific functional spaces, such as profile decompositions in the cited papers of Solimini and Bahouri, Cohen and Koch, (the latter verifies the property named Assumption 1 which, as we shown above, implies cocompactness). It is not clear yet, when there are profile decompositions with the weak, but not with the strong, remainder, but we expect that this may occur only under some special conditions, such as absence of some strong convexity. The abstract profile decompositions were not yet considered for non-reflexive spaces.

For the sake of simplicity we here consider the case when the set D consists of isometries on X, as this is the case most often studied in applications. A profile decomposition with quasi-isometric operators in the Hilbert space is given in [56], Theorem 3.1.

Let us define the general class of the set of isometric operators that yields general profile decompositions. As usual, pointwise (or strong) operator convergence in X, $g_k \overset{s}{\to} g$, is convergence $g_k x \to gx$ in X for any $x \in X$.

Definition 3.1. A set D of surjective linear isometries on a Banach space X, closed with respect to the pointwise operator convergence, is called a *dislocation* (or a *gauge*) set if any sequence g_k^{-1} with $g_k \in D$, that does not converge weakly to zero, has a pointwise convergent subsequence.

All operator sets in the cocompactness results of the previous section are known to be dislocation sets (see [56] for Euclidean rescalings, anisotropic rescalings on Carnot groups, actions of isometries on locally compact manifolds, and magnetic shifts, see [2] and [4] for inhomogeneous dilations in the Moser–Trudinger settings, and [32] for shifts in the Fourier variable.)

3.1. Hilbert space

In the general Hilbert space, the following profile decomposition holds. [56], Corollary 3.2.

Theorem 3.2. *Let D be a set of dislocations on a Hilbert space H and let u_k be a bounded sequence in H. There is a renumbered subsequence of u_k and sequences $\{g_k^{(n)}\} \subset D$, $w^{(n)} \in H$, $n \in \mathbb{N}$, $g_k^{(1)} = \mathrm{id}$, such that*

$$g_k^{(n)^{-1}} u_k \rightharpoonup w^{(n)},$$

$$g_k^{(n)^{-1}} g_k^{(m)} \rightharpoonup 0 \text{ whenever } m \neq n,$$

$$\|u_k\|^2 = \sum_n \|w^{(n)}\|^2 + \|r_k\|^2 + o(1),$$

where

$$r_k := u_k - \sum_n g_k^{(n)} w^{(n)} \overset{D}{\rightharpoonup} 0,$$

and the series in the last expression converges in H uniformly with respect to k. If, in addition, an imbedding $H \hookrightarrow Y$ is D-cocompact, the last expression vanishes in Y.

Note that $g_k^{(n)^{-1}} g_k^{(m)} \rightharpoonup 0$ if and only if $(g_k^{(m)} v, g_k^{(n)} w) \to 0$ for any $v, w \in H$, so this relation can be called asymptotic orthogonality. For the group of Euclidean shifts, the relation of asymptotic orthogonality means $|y_k^{(n)} - y_k^{(m)}| \to \infty$. More generally, for actions of isometries of a Riemannian manifold M, asymptotic orthogonality means that for some point $x_0 \in M$ (and then for any other point) the sequence $\eta_k^{(n)^{-1}} \eta_k^{(m)} x_0$ has no convergent subsequence. For the rescalings $u \mapsto 2^{rj} u(2^j(\cdot - y))$, $y \in \mathbb{R}^N$, $j \in \mathbb{Z}$, asymptotic orthogonality means

$$|j_k^{(n)} - j_k^{(m)}| + |y_k^{(n)} - y_k^{(m)}| \to \infty. \tag{3.3}$$

Analogous orthogonality condition arises on Carnot groups, with rescalings $2^{jr(\nu)} T_{\gamma^j} u \circ \eta$ involving inhomogeneous dilations and left group shifts ([56, Section 9.9]).

3.2. Banach space case

There is no generalization of Theorem 3.2 for the general Banach space, but there is one ([21]) for functional spaces cocompactly imbedded into $L^p(M, \mu)$ where (M, μ) is a measure space. Profile decompositions, that are not particular cases of Theorem 3.2 or Theorem 3.3 below, are summarized, to a great extent, in [7], which is dedicated to the case of Euclidean rescalings of functions on $\mathbb{R}^N$.

Theorem 3.3. *Let M be a measure space and let D be a set of dislocations on a reflexive Banach space X continuously imbedded into $L^p(M)$ with some $p > 1$, and assume that the operators in D are isometric in L^p and that the imbedding $X \hookrightarrow L^p$ is D-cocompact. let u_k be a bounded sequence in X. There is a renumbered subsequence of u_k and sequences $(g_k^{(n)}) \subset D$, $w^{(n)} \in X$, $n \in \mathbb{N}$, $g_k^{(1)} = \mathrm{id}$, such that*

$$g_k^{(n)^{-1}} u_k \rightharpoonup w^{(n)},$$

$$g_k^{(n)^{-1}} g_k^{(m)} \rightharpoonup 0 \text{ whenever } m \neq n,$$

$$\sum_n \|w^{(n)}\|_p^p \leq \liminf \|u_k\|_p^p,$$

$$u_k - \sum_n g_k^{(n)} w^{(n)} \overset{L^p}{\to} 0,$$

and the series in the last expression converges in L^p uniformly with respect to k. If X is dense in L^p, then L^p provides a metrization of the D-weak convergence.

As a corollary of this, with $\mathcal{D}^{1,p}(\mathbb{R}^N) \hookrightarrow L^{\frac{pN}{N-p}}(\mathbb{R}^N)$ cocompactly relative to rescalings $u \mapsto 2^{rj}u(2^j(\cdot - y))$, $y \in \mathbb{R}^N$, $j \in \mathbb{Z}$, $r = \frac{N-p}{p}$, one has the profile decomposition of Solimini [49]. The orthogonality condition in Solimini's profile decomposition is (3.3). Cocompactness is proved in [49] for the imbedding into $L^{\frac{pN}{N-p},q}(\mathbb{R}^N)$, $p < q \leq \infty$, but on the bounded subsets of $\mathcal{D}^{1,p}(\mathbb{R}^N)$ convergence in all these quasinorms is equivalent to that in $L^{\frac{pN}{N-p}}(\mathbb{R}^N)$.

Remark 3.4. In Section 9.9 of [56] Solimini's profile decomposition is generalized to Sobolev spaces $\mathcal{D}^{1,2}$ of Carnot groups. The proof of cocompactness there can be trivially extended, with suitable parameter changes, to $\mathcal{D}^{1,p}$ with general p, and then Solimini's profile decomposition for Carnot groups with anisotropic rescalings follows for any $p \in (1, \infty)$ from Theorem 3.3.

3.3. Wavelet bases and profile decompositions

Profile decompositions for a cocompact imbedding of functional spaces of $\mathbb{R}^N$, relative to rescalings, are established in [7] (following a number of earlier results surveyed there, starting with Gérard's paper [28]) in the weak remainder form (3.2) and with the asymptotic decoupling condition (3.3). Additional information about the profile decomposition is given there in form of stability estimates, namely ℓ^p -bounds on the sequence $\{\|w^{(n)}\|_X\}_{n\in\mathbb{N}}$ in the cases of Besov and Triebel–Lizorkin spaces, which indicates a possibility of a stronger remainder. There is a recent work with an expressed objective to replace the weak remainder (3.2) in the profile decomposition with a strong remainder, Palatucci & Pisante [44], which accomplished this task for the profile decomposition of [28], that is, for imbedding of the Bessel potential spaces $\dot{H}^s(\mathbb{R}^N)$, $s > 0$.

Profile decomposition of [7] follows from the following assumptions on the function spaces $X \hookrightarrow Y$ of $\mathbb{R}^N$.

1. The norms of X and Y are invariant with respect to shifts and to dilations $t^r u(t\cdot)$ with some $r \in \mathbb{R}$;
2. There exists a function $\psi \in X$ such that the set $D\psi$ is an unconditional Schauder basis on both X and Y where D is the set (2.3);
3. The imbedding $X \hookrightarrow Y$ satisfies condition (2.4) (which, as we above, implies cocompactness of the imbedding whenever X is reflexive);
4. There is a $C > 0$ such that for any sequence u_k bounded in X, with expansion $u_k = \sum_{g\in D} c_k(g)g\psi$, whose coefficients $c_k(g)$ converge, for each $g \in D$, to respective finite limits $c(g)$ as $k \to \infty$, the series $\sum_{g\in D} c(g)g\psi$ converges in X with $\|\sum_{g\in D} c(g)g\psi\|_X \leq C \liminf \|u_k\|_X$.

Conditions above (and thus the profile decomposition (3.2)) are verified in [7] for all imbeddings listed in Theorem 2.5 above.

3.4. Concentration in time-evolution problems

Profile decompositions for sequences of initial data (bounded in respective Sobolev norms) of dispersive evolution equations, such as nonlinear Schrödinger or wave equation, give rise to profile decomposition of finite energy solutions with these data. This type of concentration analysis is usually called energy-critical and involves the usual group of rescalings (see [30, 31, 27, 54]), and, for a survey, [32]). While for a linear equation construction a "time-evolved" profile decomposition in the energy space of initial data is straightforward, further technical effort is unvolved in extending such decomposition to solutions of equations with a non-linear term of critical growth. Below we quote a representative result from [8].

On the other hand, profile decomposition for sequences of data bounded in the Lebesgue norm, usually called mass-critical, involve a larger group of gauges. A profile decomposition by Terence Tao, based on the cocompactness of Strichartz imbedding in Theorem 2.12, is derived directly from Theorem 3.2). We refer to [32]) for further details.

Theorem 3.5 ([8]). *Let φ_k, ψ_k be bounded sequences in, respectively, $\mathcal{D}^{1,2}(\mathbb{R}^3)$ and $L^2(\mathbb{R}^3)$, satisfying the vanishing at infinity condition*

$$\lim_{R\to\infty}\limsup_{k\to\infty}\int_{|x|>R}(|\nabla\varphi_k|^2+|\psi_k|^2)\mathrm{d}x=0.$$

Let $u_k\in C(\mathbb{R}_t,\mathcal{D}^{1,2}(\mathbb{R}_x))\cap L^5_{\mathrm{loc}}(\mathbb{R}_t,L^{10}(\mathbb{R}^3)$ with $\partial_t u_k\in C(\mathbb{R}_t,L^2(\mathbb{R}^3_x))$ denote the solution (whose existence and uniqueness were established by Shatah and Struwe [46, 47]) of the equation

$$\Box u+|u|^4u=0$$

in $\mathbb{R}_t\times\mathbb{R}^3_x$, satisfying the initial condition $u(x,0)=\varphi_k\rightharpoonup\varphi$, $\partial_t u(x,0)=\psi_k\rightharpoonup\psi$, and let u be an analogous solution with the initial data φ,ψ. Then, for a renumbered subsequence, there exist sequences $t_k^{(n)}\in\mathbb{R}_t$, $x_k^{(n)}\in\mathbb{R}^3_x$, $\lambda_k^{(n)}>0$, and functions $W^{(n)}(x,t)$, $|||W^{(n)}(x,t)|||<\infty$, such that

$$u_k(x,t)=u(x,t)+\sum_{n=1}^{\ell}{\lambda_k^{(n)}}^{1/2}W^{(n)}(\lambda_k^{(n)}(t-t_k^{(n)}),\lambda_k^{(n)}(x-x_k^{(n)}))+w_k^{(\ell)}+r_k^{(\ell)},$$

where

$$\lim_{\ell\to\infty}\limsup_{k\to\infty}\|w_k^{(\ell)}\|_{L^5(\mathbb{R}_t,L^{10}(\mathbb{R}^3_x))}=0,$$

and

$$\lim_{\ell\to\infty}\limsup_{k\to\infty}|||r_k^{(\ell)}|||=0,$$

where

$$|||r|||=\sup_{t\in\mathbb{R}}\left(\int_{\mathbb{R}^3}(|\partial_t r(t,x)|^2+|\nabla_x r(t,x)|^2)\mathrm{d}x\right)^{1/2}+\|r\|_{L^5(\mathbb{R}_t,L^{10}(\mathbb{R}^3_x))}.$$

Moreover, the elementary concentrations in the profile decomposition are asymptotically decoupled in the sense that

$$\left|\log \frac{\lambda_k^{(m)}}{\lambda_k^{(n)}}\right| + (\lambda_k^{(m)} + \lambda_k^{(n)})|x_k^{(m)} - x_k^{(n)}| \to \infty \text{ whenever } m \neq n.$$

In addition, the energy functional (the quadratic portion of $|||\cdot|||$*) is additive with regard to the terms in the decomposition.*

3.5. Profile decompositions for the Moser–Trudinger imbedding.

Profile decompositions for the Moser–Trudinger imbedding can be derived from minor adaptations of Theorem 3.2 or Theorem 3.3. For the Trudinger–Moser inequality for radial functions on the unit disk, we have the following result. We recall that $N' = \frac{N}{N-1}$

Theorem 3.6 ([2]). *Let* $u_k \rightharpoonup 0$ *in* $H_0^1(B)$ *or let* $u_k \rightharpoonup 0$ *be a sequence of radial non-increasing functions in* $W_0^{1,N}(B)$ *for* $N \geq 3$*. There exist sequences* $s_k^{(n)} \to \infty$, *and* $w^{(n)} \in W_{0,\mathrm{rad}}^{1,N}(B)$ $n \in \mathbb{N}$, *such that for a renumbered subsequence,*

$$w^{(n)} = \mathrm{w} - \lim \left(s_k^{(n)}\right)^{-1/N'} u_k(r^{s_k^{(n)}}),$$

$$|\log(s_k^{(m)}/s_k^{(n)})| \to \infty \text{ for } n \neq m,$$

$$\sum_{n\in\mathbb{N}} \int_B |\nabla w^{(n)}|^N \mathrm{d}x \leq \limsup \int_B |\nabla u_k|^N \mathrm{d}x,$$

$$u_k - \sum_{n\in\mathbb{N}} \left(s_k^{(n)}\right)^{1/N'} w^{(n)}(r^{-s_k^{(n)}}) \to 0 \text{ in } \exp L^{N'},$$

and the series $\sum_{n\in\mathbb{N}} \left(s_k^{(n)}\right)^{1/N'} w^{(n)}(r^{s_k^{(n)}})$ *converges in* $W_0^{1,N}(B)$ *uniformly in* k.

Remark 3.7. An immediate generalization of this result to the radial subspace of $W^{1,N}(\mathbb{R}^N)$ can be stated in form of the decomposition above for $u_k - u_k(1) \in H_0^1(B)$ with vanishing of $u_k(r)$ for $r > 1$. Such decomposition is presented for $N = 2$ (with an additional assumption on the sequence and the weak remainder) in [9], which is very similar to [2] and is mentioned here only on the merit of details on vanishing of a sequence $u_k \rightharpoonup 0$ in $H_{0,\mathrm{rad}}^1$ for $r > 1$. A strongly vanishing remainder is, nonetheless, immediate.

Without the assumption of radiality, profile decompositions for the Trudinger–Moser case are known when $N = 2$. The result for the bounded domain is contained in [4]. We quote it below with a trivially refined (in view of Corollary 3.2, [56]) energy estimate. Note that the concentration profiles are always radial, even when the original sequence u_k is not. Indeed, when, for example, a concentration profile is defined as the weak limit of $j_k^{-1/2} u_k(z^{j_k})$ with $j_k \to \infty$, it is obvious that they will be symmetric with respect to discrete rotations by an arbitrarily small angle, i.e., radially symmetric.

Theorem 3.8. *Let $\Omega \subset \mathbb{R}^2$ be a bounded domain and let $u_k \rightharpoonup 0$ in $H_0^1(\Omega)$. There exist sequences $j_k^{(n)} \in \mathbb{N}$, $j_k^{(n)} \to \infty$, $z_k^{(n)} \in \Omega$, and $w^{(n)} \in W_0^{1,N}(B)$ $n \in \mathbb{N}$, such that for a renumbered subsequence,*

$$w^{(n)}(|z|) = \mathrm{w} - \lim \left(j_k^{(n)}\right)^{-1/2} u_k(z_k^{(n)} + z^{j_k^{(n)}}),$$

$$z_m \neq z_n \text{ or } |\log j_k^{(m)} - \log j_k^{(n)}| \to \infty \text{ whenever } n \neq m,$$

$$\|\nabla u_k\|_2^2 = \sum_{n\in\mathbb{N}} \int_B |\nabla w^{(n)}|^2 dx + \|\nabla r_k\|_2^2 + o(1)$$

where

$$r_k := u_k - \sum_{n\in\mathbb{N}} j_k^{(n)^{1/2}} w^{(n)}(|z - z_n|^{1/j_k^{(n)}}) \overset{D}{\rightharpoonup} 0,$$

(with the latter convergence equivalent, for bounded sequences in H_0^1 to convergence in $\exp L^2$), and the series $\sum_{n\in\mathbb{N}} j_k^{(n)^{1/2}} w^{(n)}(|z - z_n|^{1/j_k^{(n)}})$ converges in H_0^1 uniformly in k.

Recently, Bahouri, Majdoub and Masmoudi [10] announced the following result, partly extending Theorem 3.8 to $\Omega = \mathbb{R}^2$.

Theorem 3.9. *Let $u_k \rightharpoonup 0$ in $H^1_{\mathrm{rad}}(\mathbb{R}^2)$ such that*

$$\limsup_{k\to\infty} \|u_k\|_{\exp_1 L^2} > 0, \tag{3.4}$$

and

$$\lim_{R\to\infty} \limsup_{k\to\infty} \|u_k\|_{L^2(|x|>R)} = 0. \tag{3.5}$$

Then, there exists a sequence of asymptotically orthogonal rescalings $s_k^{(n)}, y^{(n)}$, and a sequence of profiles $w^{(n)}$ such that, up to a subsequence extraction, for all $\ell \geq 1$, the following asymptotic relation holds true:

$$u_n(x) = \sum_{n=1}^{\ell} s_k^{(n)^{-1/2}} w^{(n)}(|x - y^{(n)}|^{s_k^{(n)}}) + \mathrm{r}_n^{(\ell)}(x), \quad \limsup_{k\to\infty} \|\mathrm{r}_k^{(\ell)}\|_{\exp_1 L^2} \overset{\ell\to\infty}{\longrightarrow} 0. \tag{3.6}$$

Moreover,

$$\|\nabla u_k\|_{L^2}^2 = \sum_{j=1}^{\ell} \|w^{(n)'}\|_{L^2}^2 + \|\nabla \mathrm{r}_n^{(\ell)}\|_{L^2}^2 + \circ(1), \quad k \to \infty. \tag{3.7}$$

In addition to inhomogeneous dilations $u \mapsto j^{-1/2}u(z^j)$, $j \in \mathbb{Z}$ of the unit disk, the gradient norm $\|\nabla u\|_2$ on the unit disk is preserved by Möbius transformations, $u(z) \mapsto u(\frac{z-\zeta}{1-\zeta\bar{z}})$, $\zeta \in B$. This has a geometric meaning of isometries of the hyperbolic plane represented by the Poincaré disk coordinates, where $\|\nabla u\|_2^2$ represents the quadratic form of the Laplace–Beltrami operator. This puts concentration compactness relative to such "translations" into the framework of Sobolev spaces on periodic manifolds, discussed in Section 2 (and based on Lemma 9.4, [56]), with the profile decomposition given by Theorem 3.2. Note only that application of Lemma 9.4, [56], gives cocompactness of the imbedding of $H_0^1(B) \hookrightarrow L^q(B,\mu)$, $q \in (2,\infty)$, where $d\mu = \frac{4dx}{(1-r^2)^2}$ is the Riemannian measure of the hyperbolic plane in the disk coordinates, and that equivalent metrizations of D-weak convergence include not only L^q norms, but any Orlicz norms $\psi L(B,\mu)$ associated with the even convex functions ψ such that $\psi(t)/t^2 \to 0$ when $t \to 0$ and $\log\psi(t)/t^2 \to 0$ when $t \to \infty$. See [3] for details.

4. Cocompactness and profile decompositions by reduction

Condition of invariance with respect to a non-compact group is of exceptional kind, and so it is important to consider the implications of cocompactness on spaces without such invariance. In many cases it is advantageous to see a Banach space X_0 as a subspace of a larger space X, that admits a profile decomposition. This profile decomposition, reduced to sequences in X_0, may take a significantly simplified form, as we see from the examples below. Furthermore, there are situations when restriction of a cocompact imbedding to a subspace results in a compact imbedding.

Example. Consider the limit Sobolev imbedding $\mathcal{D}^{1,p}(\mathbb{R}^N) \hookrightarrow L^{\frac{pN}{N-p}}(\mathbb{R}^N)$ and let u_k be a sequence bounded in the $W^{1,p}(\mathbb{R}^N)$-norm. For any such sequence the L^p-bound implies that $t_k^{\frac{N-p}{p}} u_k(t\cdot) \rightharpoonup 0$ whenever with $t_k \to 0$, and consequently, a renamed subsequence of u_k can be written as a sum of translations $w^{(n)}(\cdot - y_k^{(n)})$ plus a sum of dilations (at variable cores y_k) with $t_k \to \infty$, which, in turn vanishes in L^q, $q \in (p, \frac{pN}{N-p})$, plus a remainder that vanishes in $L^{\frac{pN}{N-p}}$ (and is still bounded in L^p, and thus also vanishes in L^q).

In other words, by considering $W^{1,p}$ as a subspace of $\mathcal{D}^{1,p}$, we derived a profile decomposition in $W^{1,p}$, with translations as only gauges and with a remainder vanishing in L^q, $q \in (p, \frac{pN}{N-p})$.

4.1. Transitivity and interpolation

Cocompactness of imbeddings is often possible to infer by means of the following elementary arguments.

A. Let D be a set of isometries on three nested Banach spaces: $X \hookrightarrow Y$ and $Y \hookrightarrow Z$, and one of the two imbeddings is cocompact. Let $g_k u_k \rightharpoonup 0$ in X for any $g_k \in D$. If the first imbedding is cocompact, then $u_k \to 0$ in Y and thus $u_k \to 0$ in Z. If the second imbedding is cocompact, then, by continuity of the first imbedding, $g_k u_k \rightharpoonup 0$ in Y and thus $u_k \to 0$ in Z.

B. Let $X \hookrightarrow Y_0$, $X \hookrightarrow Y_1$, and assume that the first imbedding is cocompact. If the convergence in Y_0 and Y_1, for sequences bounded in X, is equivalent, then obviously, the second imbedding is cocompact as well.

C. Let $X \hookrightarrow Y_0$, $X \hookrightarrow Y_\alpha$, and assume that the first imbedding is cocompact. If for any $u \in X$, $\|u\|_{Y_\alpha} \le C\|u\|_{Y_0}^\alpha \|u\|_X^{1-\alpha}$ with some $\alpha \in (0,1)$, then X is cocompactly imbedded into Y_α. Indeed, the inequality implies that $Y_0 \cap B^X \hookrightarrow Y_\alpha \cap B^X$.

In particular, two known proofs of cocompactness of the Sobolev imbedding $\mathcal{D}^{1,p}(\mathbb{R}^N) \hookrightarrow L^{p^*}$ use the argument **C**, by first establishing cocompactness of imbedding of $\mathcal{D}^{1,p}$ into a Besov space $\dot{B}^{1-N/p,\infty;\infty}$ (Jaffard [29]) or a Marcinkiewicz space $L^{\frac{pN}{N-p},\infty}$ (Solimini [49]).

In general, one would also expect that, given a common set of gauges, interpolation of two imbeddings, $X_0 \hookrightarrow Y_0$, and $X_1 \hookrightarrow Y_1$, results in a cocompact imbedding, if one of this imbeddings is cocompact. We refer to [20] where a more specific statement is proved, under additional conditions, for functional spaces of $\mathbb{R}^N$, which is then applied to verify that subcritical imbeddings of Besov spaces

$$B^s_{p,q}(\mathbb{R}^N) \hookrightarrow B^{s_1}_{p_1,q_1}(\mathbb{R}^N), \;\; q_1 \geq q, \;\; 0 < \tfrac{1}{p} - \tfrac{1}{p_1} < \tfrac{s-s_1}{N}, \;\; s_1 \geq 0.$$

are cocompact with respect to lattice shifts $D = \{u \mapsto u(\cdot - y)\}_{y \in \mathbb{Z}^N}$. This result can be also deduced from cocompactness of homogeneous Besov spaces mentioned in Section 2, by means of the reduction method below.

4.2. Reduction to a subspace

Example. Reduction of Solimini's profile decomposition for $\mathcal{D}^{1,p}(\mathbb{R}^N) \hookrightarrow L^{\frac{pN}{N-p}}$ to the subspace of radial functions eliminates from the profile decomposition all translations and leaves only the concentrations with dilations about the origin. This can be also obtained directly from Proposition 2.3 and Theorem 3.3 above.

Example. If $\Omega \subset \mathbb{R}^N$ is a bounded domain, one obtains a profile decomposition for the imbedding $W^{1,p}_0(\Omega) \hookrightarrow L^{\frac{pN}{N-p}}(\Omega)$ by reducing Solimini's decomposition as follows. By Friedrichs inequality, the sequence u_k has a L^p -bound, which and eliminates all profiles subjected to unbounded deflations. Furthermore, there are no concentrations with $t_k \to \infty$ at cores lying outside of Ω, since the corresponding weak limits will necessarily be equal zero. For the same reason, if t_k if bounded (equivalently, with $t_k = 1$), there are no concentration terms with translations by unbounded sequences y_k. As a result, every bounded sequence in $W^{1,p}_0(\Omega)$ has a subsequence consisting of a countable sum of local concentrations $t_k^{\frac{N-p}{p}} w(t_k(\cdot - y))$, $t_k \to \infty$, and a remainder vanishing in $L^{\frac{pN}{N-p}}$. This local concentration can be easily transferred, using the exponential map, to compact Riemannian manifolds, giving rise to profile decompositions, which were introduced, under the name of global compactness by Struwe [52] for Palais–Smale sequences, (see Theorem 3.1 in [24]), although, allowing infinitely many terms and arbitrary profiles, they can be easily re-established for general sequences.

4.3. Compactness as reduced cocompactness

Reduction to subspaces may in some cases eliminate all concentrations, which makes a restriction of the cocompact imbedding to a subspace a compact imbedding. In this case the argument does not involve profile decompositions and uses only cocompactness. We give here two examples: subspaces defined by restriction of support (compactness of Sobolev imbeddings on domains thin at infinity) and subspaces defined by a compact symmetry. The following statement is simpler and more general than its partial counterparts in [56], and we bring it with a proof.

Proposition 4.1. *Let M be a complete Riemannian N-manifold, periodic relative to some subgroup G of its isometries. Let X be a reflexive Banach space cocompactly imbedded into $L^q(M)$ for some $1 < q < \infty$, and assume that $\|u \circ \eta\|_q = \|u\|_q$ for all $\eta \in G$. Assume that the imbedding $X \hookrightarrow L^q(M)$ is cocompact relative to the action of G. Let $\Omega \subset M$ be an open set such that the measure of the set $\liminf \eta_k \Omega$, $\eta_k \in G$, is zero whenever $\eta_k x_0$ has no convergent subsequence for some $x_0 \in M$. If $X(\Omega)$ is the subspace of X consisting of functions that equal zero a.e. on $M \setminus \Omega$, then the imbedding $X(\Omega) \hookrightarrow L^q(\Omega)$ is compact.*

Proof. Let $u_k \in X(\Omega)$ be a bounded sequence, and assume without loss of generality, that $u_k \rightharpoonup 0$. By assumption, if $\eta_k x_0$ has no convergent subsequence, then any weakly convergent subsequence of $u_k \circ \eta_k$ converges weakly to a function, supported, as a L^q-function, on a set of measure zero, i.e., by the imbedding, to the zero element of X. Assume now that, on a renumbered subsequence, $\eta_k \to \eta$. Then $\mathrm{w}-\lim u_k \circ \eta_k = \mathrm{w}-\lim u_k \circ \eta = 0$ in X. Since this is true for any convergent subsequence, this is true for the original η_k. We conclude then that in any case $u_k \circ \eta_k \rightharpoonup 0$, and thus, by cocompactness of the imbedding, $u_k \to 0$ in $L^q(M)$. □

The following theorem includes as the particular cases, the Strauss lemma [51] as well as its generalization to subspaces with the block-radial symmetry.

Theorem 4.2 (Compactness under coersive symmetries, [48]). *Let M be a complete Riemannian N-manifold with a transitive group G of isometries. Let X be a Banach space cocompactly imbedded into L^q for some $q \in (1, \infty)$, relative to the actions of G. Let $\Omega \subset G$ be a compact subgroup and let X_Ω be a subspace of X invariant with respect to actions of Ω. The imbedding $X_\Omega \hookrightarrow L^q(M)$ is compact if and only if for any $t > 0$ the set*

$$\{x \in M : \operatorname{diam}(\Omega x) \leq t\}$$

is bounded.

4.4. Flask spaces

Profile decomposition can be naturally extended to a class of subspaces that are not gauge-invariant. Del Pino and Felmer [22] have discovered them as Sobolev spaces of "flask domains" in $\mathbb{R}^N$. On the functional-analytic level flask spaces were defined in [56].

Definition 4.3. Let D be a set of surjective isometric operators on a Banach space X. A subspace $X_0 \subset X$ is called a D-flask subspace if the set of sequential weak limits $\mathrm{w}-\lim(DX_0)$ of sequences $\{g_k u_k,\ g_k \in D, u_k \in X_0\}_{k\in\mathbb{N}}$ remains in DX_0.

This property does not hold, in particular, if $X \hookrightarrow L^q(\mathbb{R}^N)$, D includes dilations, and $X_0 \subsetneq X$ contains a continuous function, or if D includes all translations and for any $R > 0$ the subspace $X_0 \subsetneq X$, contains a function f_R positive on some ball of radius R.

On the other hand if $X(\Omega) = W_0^{1,p}(\Omega)$ where $\Omega = (-1,1) \times \mathbb{R}^{N-1} \cup \mathbb{R}^{N-1} \times (-1,1)$, an infinite cross, and D is a group of shifts by $\mathbb{Z}^N$, then any nonzero translated weak limit of a sequence from $W_0^{1,p}(\Omega)$ will be supported on a domain that is a translate of Ω, or a translate of $(-1,1) \times \mathbb{R}^{N-1}$, or a translate of $\mathbb{R}^{N-1} \times (-1,1)$, which in any case is a subset of Ω. We have an immediate statement which is an elementary generalization of Proposition 3.5 of [56].

Theorem 4.4. *Assume that D be a group of surjective isometric operators on a Banach space X, such that the profile decomposition* (3.1) *holds true. If $X_0 \subset X$ is a D-flask subspace, then any bounded sequence in X_0 has a profile decomposition* (3.1) *with profiles $w^{(n)} \in X_0$.*

Proof. Since X_0 is a D-flask space, for each $n \in \mathbb{N}$ there exists $g_n \in D$ such that $\tilde{w}^{(n)} = g_n w^{(n)} \in X_0$. The profile decomposition (3.1) can be rewritten then with $\tilde{w}^{(n)}$ replaced with $w^{(n)}$ and $g_k^{(n)}$ replaced by $\tilde{g}_k^{(n)} = g_k^{(n)} g_n^{-1}$. □

The following sufficient condition for a D-flask set is a slightly generalized version of Remark 9.1(d) in [56].

Theorem 4.5. *Let M be a complete Riemannian manifold, and let G be a subgroup of its isometries. Let X be a reflexive Banach space continuously imbedded into $L^q(M)$, $1 < q < \infty$, let $D = \{u \mapsto u \circ \eta\}_{\eta \in G}$ and assume that $\|u \circ \eta\|_X = \|u\|_X$ for all $\eta \in G$. Let $X(\Omega)$, $\Omega \subset M$, be a subspace of all functions in X that vanish a.e. in $M \setminus \Omega$. If for any sequence $\eta_k \in M$ there exist $\eta \in G$ and a set of zero measure $Z \subset M$, such that* $\liminf \eta_k \Omega \subset \eta\Omega \cup Z$, *then $X(\Omega)$ is a D-flask set.*

References

[1] Adams, R.A., Compact imbedding theorems for quasibounded domains, Trans. Amer. Math. Soc., **148** (1970), 445–459.

[2] Adimurthi; do O., J.M.; Tintarev, K., Cocompactness and minimizers for inequalities of Hardy–Sobolev type involving N-Laplacian. NoDEA **17** (2010), 467–477.

[3] Adimurhti; Tintarev, K., On a version of Trudinger–Moser inequality with Möbius shift invariance, Calc. Var. Partial Differential Equations **39** (2010), 203–212.

[4] Adimurhti; Tintarev, C., On compactness in Trudinger–Moser inequality, Annali SNS Pisa (accepted), arXiv:1110.3647.

[5] Adimurthi; Sandeep, K., A singular Moser–Trudinger embedding and its applications, NoDEA **13** (2007), 585–603.

[6] Arioli, G.; Szulkin, A., A semilinear Schrödinger equation in the presence of a magnetic field, Arch. Ration. Mech. Anal. **170** (2003), no. 4, 277–295.

[7] Bahouri, H.; Cohen, A.; Koch, G., A General Wavelet-Based Profile Decomposition in the Critical Embedding of Function Spaces, Confluentes Matematicae **3** (2011), 387–411.

[8] Bahouri, H.; Gérard, P.; High frequency approximation of solutions to critical nonlinear wave equations Amer. J. Math. **121** (1999), 131–175.

[9] Bahouri, H.; Majdoub, M.; Masmoudi, N., On the lack of compactness in the 2D critical Sobolev embedding. J. Funct. Anal. **260** (2011), 208–252.

[10] Bahouri, H.; Majdoub, M.; Masmoudi, N., Lack of compactness in the 2D critical Sobolev embedding, the general case, C. R. Math. Acad. Sci. Paris **3**50 (2012), 177–181.

[11] Ben Ameur, J.; Description du défaut de compacité de l'injection de Sobolev sur le groupe de Heisenberg, Bull. Soc. Math. Belg. **15**, 599–624, 2008.

[12] Benci, V.; Cerami, G., Positive solutions of some nonlinear elliptic problems in exterior domains, Arch. Rat. Mech. Anal. **99** (1987), 283–300.

[13] Bennett, C.; Rudnick, K., On Lorentz–Zygmund spaces, Dissertationes Math. (Rozprawy Mat.) 175 (1980), 67 pp.

[14] Biroli, M.; Schindler, I.; Tintarev, K., Semilinear equations on Hausdorff spaces with symmetries. Rend. Accad. Naz. Sci. XL Mem. Mat. Appl. (5) **27** (2003), 175–189.

[15] Brezis, H.; Lieb, E., A relation between pointwise convergence of functions and convergence of functionals, Proc. Amer. Math. Soc. **88** (1983), 486–490.

[16] Brezis, H; Coron J.M.; Convergence of solutions of H-systems or how to blow bubbles, *Archive Rat. Mech. Anal.* **89** (1985), 21–56.

[17] Cazenave, T., Semilinear Schrödinger equations. Courant Lecture Notes in Mathematics 10, New York University, Courant Institute of Mathematical Sciences, New York; American Mathematical Society, Providence, RI, 2003. xiv+323 pp.

[18] Chabrowski, J., Weak convergence methods for semilinear elliptic equations, World Scientific Publishing Co., Inc., River Edge, NJ, 1999. xii+234 pp.

[19] Clark C., An embedding theorem for function spaces, Pacific J. Math. **19** (1966), 243–251.

[20] Cwikel, M.; Tintarev, C.; On interpolation of cocompact imbeddings, Rev. Mat. Complut. **2**6 (2013), 33–55.

[21] Cwikel M., Tintarev, C., Profile decompositions in Banach spaces (in preparation).

[22] del Pino, M., Felmer, P., Least energy solutions for elliptic equations in unbounded domains, Proc. Royal Soc. Edinburgh **126A** (1996), 195–208.

[23] DeVore, R.; Jawerth B.; Popov, V.; Compression of wavelet decompositions, American Journal of Mathematics, **114** (1992), 737–785.

[24] Druet, O.; Hebey, E.; Robert, F. Blow-up theory for elliptic PDEs in Riemannian geometry, Mathematical Notes 45, Princeton University Press, Princeton, NJ, 2004. viii+218 pp.

[25] Folland, G.B., Subelliptic estimates and function spaces on nilpotent Lie groups, Ark. Mat. **1**3 (1975), 161–207.

[26] Folland, G.B.; Stein, E.M., Estimates for the ∂b complex and analysis on the Heisenberg group, Comm. Pure Appl. Math. **27** (1974), 429–522.

[27] Isabelle Gallagher. Profile decomposition for solutions of the Navier-Stokes equations. Bull. Soc. Math. France, **129** (2001), 285–316.

[28] Gérard, P., Description du défaut de compacité de l'injection de Sobolev, ESAIM Control Optim. Calc. Var. **3** (1998), 213–233.

[29] Jaffard, S., Analysis of the lack of compactness in the critical Sobolev embeddings, J. Funct. Anal. 161 (1999) 384–396.

[30] Kenig, C.E.; Merle, F., Global well-posedness, scattering and blow-up for the energy-critical focusing non-linear wave equation, Acta Math. 201 (2008), 147–212.

[31] Keraani, S., On the blow up phenomenon of the critical nonlinear Schrödinger equation, J. Funct. Anal. 235 (2006),171–192.

[32] Killip, R., Visan, M., Nonlinear Schrödinger equations at critical regularity. Lecture notes prepared for Clay Mathematics Institute Summer School, Zürich, Switzerland, 2008.

[33] G. Kyriasis, Nonlinear approximation and interpolation spaces, Journal of Approximation Theory, 113, 110–126, 2001.

[34] Li, Yuxiang; Ruf, Bernhard, A sharp Trudinger–Moser type inequality for unbounded domains in R^n. Indiana Univ. Math. J. 57 (2008), 451–480.

[35] Lieb, E., On the lowest eigenvalue of the Laplacian for the intersection of two domains. Invent. Math. **74** (1983), 441–448.

[36] Lions P.-L., The concentration-compactness principle in the calculus of variations. The locally compact case, part 1. Ann. Inst. H. Poincaré, Analyse non linéaire **1** (1984), 109–1453

[37] Lions P.-L., The concentration-compactness principle in the calculus of variations. The locally compact case, part 2. Ann. Inst. H. Poincaré, Analyse non linéaire **1** (1984), 223–283

[38] Lions P.-L., The concentration-compactness principle in the calculus of variations. The Limit Case, Part 1, Revista Matematica Iberoamericana **1.1** (1985), 145–201.

[39] Lions P.-L., The concentration-compactness principle in the calculus of variations. The Limit Case, Part 2, Revista Matematica Iberoamericana **1.2** (1985), 45–121.

[40] Lions, P.-L., Solutions of Hartree–Fock equations for Coulomb systems, *Comm. Math. Phys.* **109** (1987), 33–97.

[41] Y. Meyer, Ondelettes et opérateurs, Hermann, 1990.

[42] J. Moser, A sharp form of an inequality by N. Trudinger, Indiana Univ. Math. J. **20** (1971) 1077–1092

[43] Maz'ya, V., Sobolev spaces with applications to elliptic partial differential equations, Second, revised and augmented edition, Grundlehren der Mathematischen Wissenschaften 342, Springer, Heidelberg, 2011. xxviii+866 pp.

[44] Giampiero Palatucci, Adriano Pisante, Improved Sobolev embeddings, profile decomposition, and concentration-compactness for fractional Sobolev spaces, arXiv: 1302.5923

[45] Schindler, I.; Tintarev, K., An abstract version of the concentration compactness principle, Rev. Mat. Complut. 15 (2002), 417–436.

[46] Shatah J., Struwe, M., Regularity results for nonlinear wave equations, Ann. of Math. 138 (1993), 503–518.

[47] Shatah J., Struwe, M., Well-posedness in the energy space for semilinear wave equations with critical growth, Internat. Math. Res. Notices **7** (1994), 303–309.

[48] Skrzypczak, L., Tintarev, C.; A geometric criterion for compactness of invariant subspaces, Arch. Math (online first, DOI 10.1007/s00013-013-0554-8).

[49] Solimini, S., A note on compactness-type properties with respect to Lorentz norms of bounded subsets of a Sobolev space, Ann. Inst. H. Poincaré Anal. Non Linéaire **12** (1995), 319–337.

[50] Schindler, I.; Tintarev, K., Semilinear subelliptic problems without compactness on Lie groups, NoDEA **11** (2004), 299–309.

[51] Strauss, W.A., Existence of solitary waves in higher dimensions. Comm. Math. Phys. **55** (1977), 149–162.

[52] Struwe, M., A global compactness result for elliptic boundary value problems involving limiting nonlinearities. Math. Z. **187** (1984), 511–517.

[53] Tao, Terence, A pseudoconformal compactification of the nonlinear Schrödinger equation and applications, New York J. Math. **15** (2009), 265–282.

[54] Tao, T., A (Concentration-)Compact Attractor for High-dimensional Non-linear Schrödinger Equations, Dynamics of PDE, **4** (2007), 1–53.

[55] Schindler, I.; Tintarev, K., Semilinear equations on fractal blowups, J. Math. Anal. Appl. **352** (2009) 57–64.

[56] Tintarev K., Fieseler K.-H., Concentration compactness: functional-analytic grounds and applications, Imperial College Press 2007.

[57] Willem, M., Minimax theorems, Progress in Nonlinear Differential Equations and their Applications 24, Birkhäuser Boston, Inc., Boston, MA, 1996. x+162 pp.

[58] Yudovich, V.I., Some estimates connected with integral operators and with solutions of elliptic equations, Sov. Math., Dokl. **2** (1961), 746–749.

Cyril Tintarev
Department of Mathematics
Uppsala University
Box 480
S-751 06 Uppsala, Sweden
e-mail: `tintarev@math.uu.se`

Concentration Analysis and Applications to PDE
Trends in Mathematics, 143–156

A Note on Non-radial Sign-changing Solutions for the Schrödinger–Poisson Problem in the Semiclassical Limit

Giusi Vaira

Abstract. We study a system of equations known as Schrödinger–Poisson problem proving that the set of sign-changing solutions has a rich structure in the semiclassical limit. Indeed we construct non-radial multi-peak solutions with an arbitrary large number of positive and negative peaks which are displaced in suitable symmetric configurations and which collapse to the same point as $\epsilon \to 0$. The proof is based on the Lyapunov–Schmidt reduction.

Mathematics Subject Classification (2010). 35B40, 35J20, 35J61, 35Q40, 35Q55.

Keywords. Schrödinger–Poisson problem, semiclassical limit, cluster solutions, sign-changing solutions, variational methods, Lyapunov–Schmidt reduction.

1. Introduction

In this survey we sketch new recent results obtained in a joint work with I. Ianni [15].

We are concerned with the existence of sign-changing solutions to the following nonlinear Schrödinger–Poisson problem

$$(\mathcal{SP}) \qquad \begin{cases} -\epsilon^2\Delta v + v + \phi v = f(v) & \text{in } \mathbb{R}^N \\ -\Delta\phi = a_N v^2 & \text{in } \mathbb{R}^N \\ \phi(x) \to 0 \ \text{ as } \ |x| \to +\infty & \end{cases}$$

where ϵ is a small and positive parameter, $f : \mathbb{R} \to \mathbb{R}$ is given, $N \geq 3$, a_N is the surface measure of the unit sphere in $\mathbb{R}^N$ and the unknowns are $v, \phi : \mathbb{R}^N \to \mathbb{R}$.

Systems like $(\mathcal{SP})$ have been object of many investigations in the last years because of their strong physical meaning. Indeed they appear in quantum mechanics models (see, e.g., [7, 8, 17]) and also in semiconductor theory [5, 6, 18, 19]. In [5, 6], for instance, they have been introduced as models describing solitary waves for nonlinear stationary equations of Schrödinger type interacting with an electrostatic field.

In this context the nonlinear term $f(v)$ simulates, as usual, the interaction between many particles, while the solution ϕ of the Poisson equation plays the role of a potential determined by the charge of the wave function itself. From another point of view, the interest on this problem stems also from the Slater approximation of the exchange term in the Hartree–Fock model, see [22]. In this framework $f(v) = v^p$ with $p = 5/3$, however, other nonlinearities have been used in different approximations.

In the paper [15] we construct, for ϵ small, non-radial solutions to $(\mathcal{SP})$ such that v is nodal, moreover v is multi-peak shaped and its peaks collapse all at a certain point (which we may assume to be 0 by the invariance by translation) as $\epsilon \to 0$ (cluster nodal solutions).

We also show that the set of nodal solutions to $(\mathcal{SP})$ has a rich structure, indeed we construct several different solutions having suitable symmetric spatial displacement of the signed peaks. We recall that D'Aprile and Wei in [12] proved the existence of positive cluster solutions to $(\mathcal{SP})$ as ϵ goes to zero.

Before stating the main results we fix the assumptions on f that we will use in the sequel and we recall some known facts.

(f1) $f \in C^{1+\sigma}_{loc}(\mathbb{R})$ with $\sigma \in (0,1)$, $f(0) = f'(0) = 0$ and $f(t) = -f(-t)$.

(f2) the problem

$$\begin{cases} \Delta w - w + f(w) = 0 & \text{in } \mathbb{R}^N \\ w > 0 & \text{in } \mathbb{R}^N \\ \lim_{|x|\to+\infty} w(x) = 0 & \\ w(0) = \max_{\mathbb{R}^N} w(x) & \end{cases} \tag{1}$$

has a unique solution w which is non degenerate, i.e., denoting by $\mathcal{L} : H^2(\mathbb{R}^N) \to L^2(\mathbb{R}^N)$ the linearized operator in w, $\mathcal{L}[u] := \Delta u - u + f'(w)u$, then

$$\text{Kernel}(\mathcal{L}) = \text{span}\left\{\frac{\partial w}{\partial x_1}, \dots, \frac{\partial w}{\partial x_N}\right\}.$$

We recall that w is a critical point of the following energy functional

$$I[w] := \frac{1}{2}\int_{\mathbb{R}^N} (|\nabla w|^2 + w^2)dx - \int_{\mathbb{R}^N} F(w)dx$$

where $F(t) = \int_0^t f(s)ds$.

By the well-known result of Gidas, Ni and Nirenberg ([14]), w is radially symmetric and strictly decreasing in $r = |x|$. Moreover, by classical regularity results, the following asymptotic behaviors hold:

$$w(r), w''(r) = A_N r^{-\frac{N-1}{2}} e^{-r}\left(1 + O\left(\frac{1}{r}\right)\right), \tag{2}$$

$$w'(r) = -A_N r^{-\frac{N-1}{2}} e^{-r}\left(1 + O\left(\frac{1}{r}\right)\right), \tag{3}$$

where $A_N > 0$ is a suitable positive constant.

The class of nonlinearities f satisfying (f1)–(f2) includes, and it is not restricted to, the homogeneous nonlinearity $f(v) = |v|^{p-1}v$ with $p \in (1, \frac{N+2}{N-2})$.

In this paper the dimension N is chosen in the interval $[3,6]$. Under this assumption it is well known that the system $(\mathcal{SP})$ can be reduced into a single equation. Indeed a simple application of the Lax–Milgram theorem ensures the existence of a unique solution of the second equation of $(\mathcal{SP})$, namely the following result holds:

Lemma 1.1. *Let $N \in [3,6]$. For every $f \in L^{\frac{2N}{N+2}}(\mathbb{R}^N)$ there exists a unique solution $\phi[f]$ in $D^{1,2}(\mathbb{R}^N)$ of the equation $-\Delta\phi = a_N f$. Moreover the following representation formula holds:*

$$\phi[f](x) := \int_{\mathbb{R}^N} \frac{f}{|x-y|^{N-2}} dy.$$

Furthermore the functional $G : H^1(\mathbb{R}^N) \to \mathbb{R}$

$$G(u) := \int_{\mathbb{R}^N} \phi[u^2]u^2 dx$$

is C^1 and $G'(u)[v] = 4\int_{\mathbb{R}^N} \phi[u^2]uv dx$.

By Lemma 1.1 we reduce to study the following nonlinear scalar equation in $H^1(\mathbb{R}^N)$

$$-\epsilon^2\Delta v + v + v\phi[v^2] = f(v) \qquad \text{in } \mathbb{R}^N. \tag{4}$$

We also recall that the solutions of (4) are critical points of the C^2-functional $J_\epsilon : H^1(\mathbb{R}^N) \to \mathbb{R}$ defined as

$$J_\epsilon[v] = \frac{1}{2}\int_{\mathbb{R}^N} \left(\epsilon^2|\nabla v|^2 + v^2\right) dx - \int_{\mathbb{R}^N} F(v)dx + \frac{1}{4}\int_{\mathbb{R}^N} \phi[v^2](x)v^2(x)dx.$$

We can now state our results. Our first theorem is about the existence of nodal solutions whose form consists of one positive peak centered in 0 surrounded by k negative peaks located near the vertices of a regular polygon, with the number k sufficiently large.

Theorem 1.2. *Let (f1)and (f2) hold and let $N \in [3,6]$. Fix $k \geq 7$ and let $Q_1, \ldots, Q_k \in \mathbb{R}^2$ be the vertices of a two-dimensional convex regular polygon centered at 0. Then there exists $\epsilon_0 > 0$ such that for any $\epsilon \in (0, \epsilon_0)$, there is $r_\epsilon > 0$ and a sign-changing solution $v_\epsilon \in H^1(\mathbb{R}^N)$ to (4) of the form*

$$v_\epsilon(x) = w\left(\frac{x}{\epsilon}\right) - \sum_{i=1}^{k} w\left(\frac{x - P_{i\epsilon}}{\epsilon}\right) + h.o.t., \quad \text{as } \epsilon \to 0$$

uniformly for $x \in \mathbb{R}^N$.

Here $P_{i\epsilon} := (r_\epsilon Q_i, \mathbf{0}) \in \mathbb{R}^N$, $i = 1, \ldots, k$ and $r_\epsilon \to 0$ as $\epsilon \to 0$. Moreover

$$\lim_{\epsilon \to 0} \frac{r_\epsilon}{\epsilon \log \frac{1}{\epsilon^2}} = C$$

for some $C > 0$.

More in general we can prove the following result:

Theorem 1.3. *Let $Q_1, \dots, Q_k \in \mathbb{R}^h$ $(2 \le h \le N)$ be the vertices of a convex regular polytope in $\mathbb{R}^h$ centered at 0 and having radius 1 and side s.*

Assume $s \le 1$, $h > 2$ or $s < 1$, $h = 2$. Then there exists $\epsilon_0 > 0$ such that for any $\epsilon \in (0, \epsilon_0)$, there is a sign-changing solution v_ϵ to (4) with one positive peak centered in 0 and k negative peaks centered at points $P_{i\epsilon} := (r_\epsilon Q_i, \mathbf{0}) \in \mathbb{R}^N$, $i = 1, \dots, k$, such that $r_\epsilon \to 0$ as $\epsilon \to 0$. Moreover

$$\lim_{\epsilon \to 0} \frac{r_\epsilon}{\epsilon \log \frac{1}{\epsilon^2}} = C$$

for some $C > 0$.

We may also construct solutions having more than two nodal regions, alternating (suitable) convex regular polygons having respectively negative and positive peaks on their vertices. For instance we construct a solution with a positive peak centered in zero surrounded by k negative peaks and $h = km$ positive peaks which are respectively on the vertices of two nested polygons, but of course this construction could be generalized to more than two nested polygons.

Theorem 1.4. *Fix $k, m \in \mathbb{N}$ such that*

$$k > \min \left\{ 6, \frac{1}{m} \frac{\pi}{\arcsin 1/4} \right\} \tag{5}$$

and let $Q_1, \dots, Q_{km} \in \mathbb{R}^2$ be the vertices of a two-dimensional convex regular polygon centered at 0. Then there exists $\epsilon_0 > 0$ such that for any $\epsilon \in (0, \epsilon_0)$ there is $r_\epsilon > 0$ and a sign-changing solution $v_\epsilon \in H^1(\mathbb{R}^N)$ to (4) of the form

$$v_\epsilon(x) = w\left(\frac{x}{\epsilon}\right) - \sum_{i=1}^{k} w\left(\frac{x - P_{i\epsilon}}{\epsilon}\right) + \sum_{i=1}^{km} w\left(\frac{x - P_{k+i_\epsilon}}{\epsilon}\right) + h.o.t., \quad as\ \epsilon \to 0$$

uniformly for $x \in \mathbb{R}^N$. Here

$$\mathbb{R}^N \ni P_{j_\epsilon} := \begin{cases} (r_\epsilon Q_{(j-1)m+1}, \mathbf{0}) & j = 1, \dots, k \\ (2 r_\epsilon Q_{j-k}, \mathbf{0}) & j = k+1, \dots, k + km \end{cases}$$

and $r_\epsilon \to 0$ as $\epsilon \to 0$. Moreover

$$\lim_{\epsilon \to 0} \frac{r_\epsilon}{\epsilon \log \frac{1}{\epsilon^2}} = C$$

for some $C > 0$.

At the end, we construct a solution having $k = m_1 q$ negative and $h = m_2 q$ positive peaks alternated on the vertices of $d = \max\{m_1, m_2\}$ suitable nested regular polygons, with the same sign on corresponding vertices, precisely

Theorem 1.5. *Let (f1) and (f2) hold and let $N \in [3, 6]$. Fix $q \in \mathbb{N}$, $q \ge 2$ and $(m_1, m_2) \in \mathbb{N} \times \mathbb{N} \setminus \{(1,1)\}$. Let $Q_1, \dots, Q_{2q} \in \mathbb{R}^2$ be the vertices of a two-dimensional convex regular polygon centered at 0. Then there exists $\epsilon_0 > 0$ such* *that for any* $\epsilon \in (0, \epsilon_0)$ *there is* $r_\epsilon > 0$ *and a sign-changing solution* $v_\epsilon \in H^1(\mathbb{R}^N)$ *to* (4) *of the form*

$$v_\epsilon(x) = \sum_{i=1}^{2q} (-1)^i \left(\sum_{j=1}^{m_{I(i)}} w\left(\frac{x - P_{ij_\epsilon}}{\epsilon} \right) \right) + h.o.t., \quad as\ \epsilon \to 0$$

uniformly for $x \in \mathbb{R}^N$.
Here $I(i) := \frac{(-1)^i+3}{2}$ *and the points* $P_{ij_\epsilon} := \left(\left(1 + (j-1)\sin\frac{\pi}{2q}\right) r_\epsilon Q_i, \mathbf{0} \right) \in \mathbb{R}^N$, $j = 1, \dots, m_{I(i)}$, $i = 1, \dots, 2q$.

Moreover $r_\epsilon \to 0$ *as* $\epsilon \to 0$ *and*

$$\lim_{\epsilon \to 0} \frac{r_\epsilon}{\epsilon \log \frac{1}{\epsilon^2}} = C$$

for some $C > 0$.

The proof of our results is based on the well-known Lyapunov–Schmidt reduction procedure (see [2]). In particular, in order to deal with nodal clustered solutions, we perform the reduction in suitable symmetric settings in the spirit of [11]. We outline here the main ideas.

First our approximate solutions are constructed as the sum (with sign) of suitably rescaled w centered at distinct points $P_i \in \mathbb{R}^N$ such that $P_i \to 0$ as $\epsilon \to 0$.

This choice is the most natural. In fact if v is a solution of (4) and $P \in \mathbb{R}^N$ then $v_\epsilon(x) := v(\epsilon x + P)$ solves the equation $-\Delta u + u + \epsilon^2 \phi[u^2]u = f(u)$ which, since $\epsilon \to 0$, can be approximated by problem (1). Hence it is quite natural to take $v \sim \pm w(\frac{x-P}{\epsilon})$ as a solution of (4) for ϵ small.

Moreover, if we take several different fixed points $P_i \in \mathbb{R}^N$, then $v \sim \sum_i \pm w(\frac{x-P_i}{\epsilon})$ is still a good approximation of a solution v of (4) for ϵ small enough, in spite of the presence of nonlinear terms in the equation. The reason is that, thanks to the exponential decay of w, the interactions among peaks centered at different fixed points becomes negligible when $\epsilon \to 0$.

In our case however we are looking for clustered solutions, namely the points $P_i \to 0$ as $\epsilon \to 0$. This means that the interactions among peaks play a role.

Anyway by locating the peaks in suitable symmetric configurations, still we will be able to find a solution of the desired form.

Indeed we recall that the Lyapunov–Schmidt reduction method reduces the problem to find a critical point for a functional defined on a finite-dimensional space (reduced functional). In our case the reduced functional, up to a positive constant, has the form

$$M_\epsilon[(P_1, \dots, P_\ell)] = \epsilon^2 \sum_{i \neq j} \frac{1}{|\frac{P_i - P_j}{\epsilon}|^{N-2}} - \sum_{i \neq j, \lambda_i = \lambda_j} w\left(\frac{P_i - P_j}{\epsilon} \right)$$
$$+ \sum_{i \neq j, \lambda_i = -\lambda_j} w\left(\frac{P_i - P_j}{\epsilon} \right) + \ h.o.t.,$$

where $\lambda_i = \pm 1$ according to the sign of each peak and the unknowns P_i determine the location of the peaks. We notice that it consists of three main terms: the first term depends on the Poisson potential effect, the second term is due to the interplay between the peaks of the same sign and has a repulsive effect, the third term is due to the interaction between peaks of opposite sign and has an attractive effect.

Observe that the first term $\epsilon^2 \sum_{i\neq j} \frac{1}{|\frac{P_i - P_j}{\epsilon}|^{N-2}}$ increases when the points P_i are close to zero, while using the exponential decay of w, the interaction term $-\sum_{i\neq j, \lambda_i = \lambda_j} w\left(\frac{P_i - P_j}{\epsilon}\right)$ increases when the mutual distance between the points P_i is big. Hence, if we restrict the functional to suitable symmetric configurations in which the peaks having opposite sign are kept away from each other, then the mutual interaction between opposite peaks, i.e., the third term $+\sum_{i\neq j, \lambda_i = -\lambda_j} w\left(\frac{P_i - P_j}{\epsilon}\right)$, becomes negligible and so we can easily conclude that the equilibrium is achieved for a suitable (symmetric) configuration of the points P_i, which is a local maximum for the functional M_ϵ, namely we have produced a sign-changing cluster solution for the problem $(\mathcal{SP})$.

We remark that one can find solutions with analogous symmetric configurations as in all the previous results, also in the case of the scalar Schrödinger equation

$$-\epsilon^2 \Delta v + V(|x|)v = |v|^{p-1}v \qquad x \in \mathbb{R}^N$$

in presence of a radially symmetric potential V with a local maximum in 0 (see also [11]).

Notations

Before going on we establish some notations. Let us denote by $H^1(\mathbb{R}^N)$ the usual Sobolev space endowed with scalar product and norm

$$(u,v)_\epsilon = \int_{\mathbb{R}^N} \left(\epsilon^2 \nabla u \nabla v + uv\right) dx; \qquad \|u\|^2 := \int_{\mathbb{R}^N} \left(\epsilon^2 |\nabla u|^2 + u^2\right) dx$$

and by $D^{1,2}(\mathbb{R}^N)$ the completition of the space $C_c^\infty(\mathbb{R}^N)$ with respect to the norm

$$\|u\|_{D^{1,2}} := \left(\int_{\mathbb{R}^N} \epsilon^2 |\nabla u|^2 \, dx\right)^{1/2}.$$

Moreover let $L^p(\mathbb{R}^N)$ the usual Lebesgue space endowed with the norm

$$|u|_p := \left(\int_{\mathbb{R}^N} |u|^p \, dx\right)^{\frac{1}{p}} \quad p \in [1,\infty) \qquad \|u\|_\infty = \sup_{x\in\mathbb{R}^N} |u(x)|.$$

In particular, let us denote by $\langle \cdot, \cdot \rangle$ the usual scalar product in $L^2(\mathbb{R}^N)$, namely

$$\langle u, v \rangle := \int_{\mathbb{R}^N} uv \, dx.$$

2. The finite-dimensional reduction

First of all we have to define the set of approximating solutions around which we do the Lyapunov–Schmidt. Precisely, for any $\ell \in \mathbb{N}$ we define

$$\Gamma_\epsilon := \left\{ \mathbf{P} = (P_1, \dots, P_\ell) \in \mathbb{R}^{N\ell} : \ \beta^2 \epsilon \log \frac{1}{\epsilon^2} < |P_i - P_j| < \epsilon \left(\log \frac{1}{\epsilon^2} \right)^2 \ \text{ for } i \neq j \right\}$$

where $\beta \in (\sigma, 1)$ is sufficiently close to 1. Let $\mathbf{P} \in \bar{\Gamma}_\epsilon$ and set $w_{P_i}(x) = w(\frac{x-P_i}{\epsilon})$, $i = 1, \dots, \ell$. We look for solutions of (4) of the form

$$v_\epsilon(x) := w_{\mathbf{P}}(x) + \psi_\epsilon(x)$$

where ψ_ϵ will be a remainder term belonging to a suitable space and the approximating solution $w_{\mathbf{P}}$ is of the form

$$w_{\mathbf{P}}(x) = \sum_{i=1}^{\ell} \lambda_i w_{P_i}(x)$$

with $\lambda_i = \pm 1$ according to the sign of each peak.

In particular we will reduce ourselves to symmetric configurations, finding solutions v_ϵ with some symmetric properties. Here we show that $\phi[v_\epsilon^2]$ preserves the same symmetry property. Indeed, let G be a group of symmetries of $\mathbb{R}^N$ and let $g \in G$. For $u : \mathbb{R}^N \to \mathbb{R}$ we set

$$(T_g u)(x) = u(gx). \tag{6}$$

Let

$$X := \left\{ u \in H^1(\mathbb{R}^N) \ : \ T_g u = u, \ g \in G \right\}$$

and

$$Y := \left\{ \phi \in D^{1,2}(\mathbb{R}^N) \ : \ T_g \phi = \phi, \ g \in G \right\}.$$

We remark that X, Y are the subspaces of H^1 and $D^{1,2}$ respectively invariant under the action (6). Indeed, if $u \in X$ then $\phi[u^2] \in Y$.

2.1. Error evaluation

Since we look for a solution near $w_{\mathbf{P}}$, a key step is to evaluate $\mathcal{S}(w_{\mathbf{P}})$ where

$$\mathcal{S}_\epsilon(v) := \epsilon^2 \Delta v - v + f(v) - \phi[v^2] v.$$

What we can prove is the following result (for the proof see for instance [12]):

Lemma 2.1. *Let $\beta \in (\sigma, 1)$. There exists a constant $C > 0$ such that for every $\epsilon > 0$ and $\mathbf{P} = (P_1, \dots, P_l) \in \mathbb{R}^{N\ell}$ with $|P_i - P_j| \geq 2\beta^2 \epsilon \log \frac{1}{\epsilon}$, for $i \neq j$:*

$$|S_\epsilon[w_{\mathbf{P}}]| \leq C \epsilon^{\beta^2(\beta^2+\sigma)} \sum_{i=1}^{l} w_{P_i}^{1-\beta^2}.$$

2.2. Linear part

Let $\mu > 0$ a sufficiently small number. We introduce the following weighted norm:

$$\|v\|_{*,\mathbf{P}} := \sup_{x\in\mathbb{R}^N} e^{\mu \min_{i=1,\dots,\ell} \frac{|x-P_i|}{\epsilon}} |v(x)|,$$

and the spaces

$$C_{*,\mathbf{P}} = \{v \in C(\mathbb{R}^N) \ : \ \|v\|_{*,\mathbf{P}} < \infty\}, \qquad H^2_{*,\mathbf{P}} = H^2(\mathbb{R}^N) \cap C_{*,\mathbf{P}}.$$

Let $\mathbf{P} \in \Gamma_\epsilon$. We define

$$Z_{P_i,j} = f'(w_{P_i}) \frac{\partial w_{P_i}}{\partial x_j} \qquad j \in \{1,\dots,N\}.$$

We consider the following linear problem:

Taken $\mathbf{P} \in \Gamma_\epsilon$ and given $h \in C_{*,\mathbf{P}}$ find a function v and constants $c_{i,j}$ satisfying

$$\begin{cases} \mathcal{L}_\mathbf{P}[v] = h + \sum_{i,j} c_{i,j} Z_{P_i,j} \\ v \in H^2_{*,\mathbf{P}}, \qquad \langle v, Z_{P_i,j}\rangle = 0, \ i = 1,\dots,\ell, \ j = 1,\dots,N. \end{cases} \tag{7}$$

where

$$\mathcal{L}_\mathbf{P}[v] := \epsilon^2 \Delta v - v + f'(w_\mathbf{P})v - \phi[w_\mathbf{P}]v - 2\phi[w_\mathbf{P} v].$$

There hold:

Lemma 2.2. *There exists $C > 0$ such that, provided ϵ is sufficiently small, if $\mathbf{P} \in \bar{\Gamma}_\epsilon$ and $(v, c_{i,j}, h)$ satisfies (7) the following holds*

$$\|v\|_{*,\mathbf{P}} \le C \|h\|_{*,\mathbf{P}}.$$

Lemma 2.3. *For $\epsilon > 0$ sufficiently small, for $\mathbf{P} \in \bar{\Gamma}_\epsilon$ and $h \in C_{*,\mathbf{P}}$ there exists a unique pair $(v, c_{i,j}) \in H^2_{*,\mathbf{P}} \times \mathbb{R}^{N\ell}$ solving (7). Furthermore by Lemma 2.2*

$$\|v\|_{*,\mathbf{P}} \le C \|h\|_{*,\mathbf{P}}.$$

2.3. Solution of the auxiliary equation

We are in the position to solve the following nonlinear system with the unknowns $(\psi, c_{i,j}) \in H^2_{*,\mathbf{P}} \times \mathbb{R}^{N\ell}$:

$$\begin{cases} S_\epsilon[w_\mathbf{P} + \psi] = \sum_{i,j} c_{i,j} Z_{P_i,j}, \\ \psi \in H^2_{*,\mathbf{P}}, \ \langle \psi, Z_{P_i,j}\rangle = 0, \quad i = 1,\dots,\ell, \ j = 1,\dots,N. \end{cases} \tag{8}$$

where $\mathbf{P} \in \bar{\Gamma}_\epsilon$ and

$$S_\epsilon[v] = \epsilon^2 \Delta v - v + f(v) - \phi[v^2]v.$$

It holds:

Lemma 2.4. *Fix $\tau = \beta^4(1+\sigma)$. Provided $\epsilon > 0$ sufficiently small, for every $\mathbf{P} \in \bar{\Gamma}_\epsilon$ there is a unique pair $(\psi_\mathbf{P}, c_{i,j}(\mathbf{P})) \in H^2_{*,\mathbf{P}} \times \mathbb{R}^{N\ell}$ which solve (8). Moreover*

$$\|\psi_\mathbf{P}\|_\infty < \epsilon^\tau; \quad (\psi_\mathbf{P}, \psi_\mathbf{P})_\epsilon \le \epsilon^{N+2\tau} \tag{9}$$

and the maps $\mathbf{P} \in \bar{\Gamma}_\epsilon \longmapsto \psi_\mathbf{P} \in H^1(\mathbb{R}^N)$ and $\mathbf{P} \longmapsto c_{i,j}(\mathbf{P}) \in \mathbb{R}$ are continuous.

Lemma 2.5. *For $\epsilon > 0$ sufficiently small the map $\mathbf{P} \in \bar{\Gamma}_\epsilon \longmapsto \psi_{\mathbf{P}} \in H^1$ constructed in Lemma 2.4 is C^1.*

2.4. The bifurcation equation

For $\epsilon > 0$ sufficiently small we define the reduced functional

$$M_\epsilon : \bar{\Gamma}_\epsilon \to \mathbb{R},\ M_\epsilon[\mathbf{P}] := \epsilon^{-N} J_\epsilon[w_{\mathbf{P}} + \psi_{\mathbf{P}}] - \ell I[w] - \epsilon^2 C_1,$$

where $\psi_{\mathbf{P}}$ has been constructed in Lemma 2.4 and C_1 is some known constant.

Next proposition contains the key expansion of M_ϵ

Proposition 2.6. *For $\epsilon > 0$ sufficiently small the following holds:*

$$M_\epsilon[\mathbf{P}] = -(\gamma_0 + o(1)) \sum_{i \neq j} \lambda_i \lambda_j w\left(\frac{P_i - P_j}{\epsilon}\right) + \epsilon^2 (C_2 + o(1)) \sum_{i \neq j} \frac{1}{|\frac{P_i - P_j}{\epsilon}|^{N-2}} + O(\epsilon^{2\tau}),$$

uniformly for $\mathbf{P} \in \bar{\Gamma}_\epsilon$, where $\tau = \beta^4(1+\sigma)$ is given by Lemma 2.4 and γ_0, C_2 are some positive known constants.

We recall the following result (whose proof can be found for instance in [12]) that will be useful in the next sections in order to find a critical point (a maximum) of M_ϵ under symmetry assumptions

Lemma 2.7. *Fix a positive constant $C > 0$ and consider the function*

$$\alpha_{\epsilon,C}(\rho) := -\gamma_0 w(\rho) + C\epsilon^2 \frac{1}{\rho^{N-2}}, \quad \rho \geq \beta^2 \log \frac{1}{\epsilon^2},$$

where γ_0 is a known positive constant. Then for $\epsilon > 0$ small enough, $\alpha_{\epsilon,C}$ has a unique maximum point ρ_ϵ. Moreover we have

$$\rho_\epsilon = \log \frac{1}{\epsilon^2} + \frac{N-1}{2} \log\log \frac{1}{\epsilon^2} + o\left(\log\log \frac{1}{\epsilon^2}\right)$$

and

$$\alpha_{\epsilon,C}(\rho_\epsilon) = C\epsilon^2 \frac{1}{\left(\log \frac{1}{\epsilon^2}\right)^{N-2}} (1 + o(1)).$$

3. Proof of Theorem 1.2

Here we sketch the proof of Theorem 1.2. For the others one can proceed in a very similar way.

For every $x \in \mathbb{R}^N$ we set $x = (x_1, \ldots, x_N) = (x_1, x_2, x') = (z, x')$, where $z \in \mathbb{C}$.

In this section we prove the existence of a cluster solution to (4) having a positive bump in 0 and k negative bumps at the vertices of a regular polygon centered in 0. Precisely we make the following *ansatz*

$$v = w_0 - \sum_{i=1}^{k} w_{P_i} + \psi_\epsilon$$

where $P_i := rQ_i \in \mathbb{R}^N$, $Q_i := (e^{2\pi\sqrt{-1}(i-1)/k}, \mathbf{0}) = \left(\cos\frac{2\pi(i-1)}{k}, \sin\frac{2\pi(i-1)}{k}, \mathbf{0}\right) \in \mathbb{R}^N$, $i = 1, \ldots, k$, and $r > 0$ is such that $\mathbf{P}_r := (0, P_1, \ldots, P_k) = (0, rQ_i, \ldots, rQ_k) \in \Gamma_\epsilon(\subset \mathbb{R}^{(k+1)N})$. Moreover we look for a solution v satisfying the following symmetry property

$$v(x_1, x_2, x') = v(z, |x'|) = v(ze^{2\pi\sqrt{-1}/k}, |x'|).$$

This translates into restricting to work into the following Sobolev space of symmetric functions

$$X := \{v \in H^1(\mathbb{R}^N) : v(x_1, x_2, x') = v(z, |x'|) = v(ze^{2\pi\sqrt{-1}/k}, |x'|)\}.$$

Hence, for every $r > 0$ such that $\mathbf{P}_r \in \Gamma_\epsilon$, we set

$$H^2_{*,r,s} := H^2_{*,\mathbf{P}_r} \cap X, \quad C_{*,r,s} := C_{*,\mathbf{P}_r} \cap X,$$

and proceeding as before finding for ϵ small enough a unique solution

$$(\psi_{\mathbf{P}_r}, c_{ij}(\mathbf{P}_r)) \in H^2_{*,r,s} \times \mathbb{R}^{N\ell}$$

to problem (8) (see Lemma 2.4).

To conclude the proof it is sufficient to find for ϵ small, a critical point $\mathbf{P}_r$ of the reduced functional M_ϵ. Indeed one can prove the following (see for instance [12])

Lemma 3.1. *Let $\mathbf{P}_r \in \Gamma_\epsilon$ be an interior maximum point for M_ϵ. Then provided that $\epsilon > 0$ is small enough, the corresponding function $v_\epsilon := w_{\mathbf{P}_r} + \psi_{\mathbf{P}_r} \in X$ is a critical point of J_ϵ in X.*

Hence, if we denote by G the group of the rotation matrix in $\mathbb{R}^{N-2}$, and for every $i \in \mathbb{N}$ and $g \in G$ we define

$$T_{i,g} : \mathbb{R}^N \to \mathbb{R}, \quad T_{i,g}(y) = T_{i,g}(z, y') = (ze^{2\pi i\sqrt{-1}/k}, gy'),$$

then the functional J_ϵ is invariant under the action of the group $\{T_{i,g} : i \in \mathbb{N}, g \in G\}$, namely $J_\epsilon(u(T_{i,g}(x))) = J_\epsilon(u(x))$. Moreover $X = \{u \in H^1(\mathbb{R}^N) : u(T_{i,g}(x)) = u(x)\}$. So the principle of symmetric criticability ensures that v_ϵ in Lemma 3.1 is also a critical point of J_ϵ and, consequently, a solution of (4).

The remaining part of the section is then devoted to find an interior maximum point of the reduced functional. Let us observe that, under our ansatz, M_ϵ is a one-variable function and, thanks to Proposition 2.6 and to the assumption $k \geq 7$, it reduces to the following

Proposition 3.2. *For $\epsilon > 0$ sufficiently small*

$$M_\epsilon[r] = (2k + o(1))\alpha_{\epsilon,C_k}\left(\frac{r}{\epsilon}\beta_k\right) + O(\epsilon^{2\tau}) \tag{10}$$

uniformly for $r > 0$ such that $\mathbf{P}_r \in \bar{\Gamma}_\epsilon$, where $\beta_k := 2\sin\frac{\pi}{k}$, C_k is a positive constant and α_{ϵ,C_k} is the function defined in Lemma 2.7.

Proof. For r such that $\mathbf{P}_r \in \bar{\Gamma}_\epsilon$ the reduced functional becomes

$$\begin{aligned} M_\epsilon[r] = & -(\gamma_0 + o(1)) \left(-2kw\left(\frac{r}{\epsilon}\right) + \sum_{i\neq j} w\left(\frac{r}{\epsilon}|Q_i - Q_j|\right)\right) \\ & + \epsilon^2 (C_2 + o(1)) \frac{1}{\left|\frac{r}{\epsilon}\right|^{N-2}} \left(2k + \sum_{i\neq j} \frac{1}{|Q_i - Q_j|^{N-2}}\right) + O(\epsilon^{2\tau}) \\ = & -k(\gamma_0 + o(1)) \left(-2w\left(\frac{r}{\epsilon}\right) + \sum_{i=2}^{k} w\left(\frac{r}{\epsilon}|Q_1 - Q_i|\right)\right) \\ & + \epsilon^2 k (C_2 + o(1)) \frac{1}{\left|\frac{r}{\epsilon}\right|^{N-2}} \left(2 + \sum_{i=2}^{k} \frac{1}{|Q_1 - Q_i|^{N-2}}\right) + O(\epsilon^{2\tau}) \\ = & -k(\gamma_0 + o(1)) \left(-2w\left(\frac{r}{\epsilon}\right) + 2w(\frac{r}{\epsilon}\beta_k) + \sum_{i=3}^{k-1} w\left(\frac{r}{\epsilon}\beta_k^i\right)\right) \\ & + \epsilon^2 k (C_2 + o(1)) \frac{1}{\left|\frac{r}{\epsilon}\right|^{N-2}} \left(2 + 2\frac{1}{\beta_k^{N-2}} + \sum_{i=3}^{k-1} \frac{1}{(\beta_k^i)^{N-2}}\right) + O(\epsilon^{2\tau}), \end{aligned}$$

where we set

$$\beta_k := |Q_2 - Q_1| = |Q_k - Q_1| = 2\sin\frac{\pi}{k}$$

and

$$\beta_k^i := |Q_i - Q_1| = \sqrt{2}\sqrt{1 - \cos\frac{2\pi(i-1)}{k}}, \quad i = 3, \dots, k-1.$$

Observe that by our choice

$$\beta_k < \beta_k^i, \quad i = 3, \dots, k-1$$

hence, from (2), it follows that

$$w\left(\frac{r}{\epsilon}\beta_k^i\right) = o\left(w\left(\frac{r}{\epsilon}\beta_k\right)\right), \quad i = 3, \dots, k-1, \quad \text{as } \frac{r}{\epsilon} \to +\infty.$$

Moreover $\beta_k < 1$ because $k \geq 7$, hence we also have

$$w\left(\frac{r}{\epsilon}\right) = o\left(w\left(\frac{r}{\epsilon}\beta_k\right)\right), \quad \text{as } \frac{r}{\epsilon} \to +\infty.$$

As a consequence the reduced functional becomes

$$M_\epsilon[r] = -(2k\gamma_0 + o(1)) w\left(\frac{r}{\epsilon}\beta_k\right) + \epsilon^2 \left(2kC_k + o(1)\right) \frac{1}{\left|\frac{r}{\epsilon}\beta_k\right|^{N-2}} + O(\epsilon^{2\tau}),$$

where $C_k := C_2 \left(1 + \beta_k^{N-2} + \frac{1}{2}\sum_{i=3}^{k-1} \frac{\beta_k^{N-2}}{(\beta_k^i)^{N-2}}\right)$. □

Observe now that $\mathbf{P}_r \in \Gamma_\epsilon$ if and only if

$$\begin{cases} \beta_k^2 \epsilon \log\frac{1}{\epsilon^2} < r < \epsilon\left(\log\frac{1}{\epsilon^2}\right)^2 \\ \beta_k^2 \epsilon \log\frac{1}{\epsilon^2} < r|Q_i - Q_j| < \epsilon\left(\log\frac{1}{\epsilon^2}\right)^2 \quad i \neq j, \end{cases}$$

hence, since $\beta_k < 1$ and $\beta_k \leq |Q_i - Q_j| \leq 2$ for $i \neq j$, it is easy to show that $\mathbf{P}_r \in \Gamma_\epsilon$ for $r \in R_\epsilon$ where

$$R_\epsilon := \left\{ r > 0 \; : \; \beta_k \epsilon \log \frac{1}{\epsilon^2} < r < \frac{1}{2}\epsilon \left(\log \frac{1}{\epsilon^2} \right)^2 \right\}.$$

Finally next result gives an interior maximum point r for M_ϵ

Proposition 3.3. *For $\epsilon > 0$ sufficiently small, the following maximization problem*

$$\max\{M_\epsilon[r] \; : \; r \in \bar{R}_\epsilon\}$$

has a solution $r_\epsilon \in R_\epsilon$. Furthermore

$$\lim_{\epsilon \to 0} \frac{r_\epsilon \beta_k}{\epsilon \log \frac{1}{\epsilon^2}} = 1.$$

Proof. Since M_ϵ is continuous in r, there exists $r_\epsilon \in \bar{R}_\epsilon$ such that

$$M_\epsilon[r_\epsilon] = \max_{r \in \bar{R}_\epsilon} M_\epsilon[r].$$

We claim that $r_\epsilon \in R_\epsilon$. We prove this by energy comparison. We first obtain a lower bound for $M_\epsilon[r_\epsilon]$. Let us choose $s_\epsilon := \frac{\epsilon \rho_\epsilon}{\beta_k}$, where $\rho_\epsilon > 0$ is given in Lemma 2.7. It is easy to see that s_ϵ belongs to R_ϵ. Indeed, by Lemma, 2.7 $\rho_\epsilon > \beta^2 \log \frac{1}{\epsilon^2}$ and, for ϵ small, $\rho_\epsilon < \frac{1}{2}\beta_k \left(\log \frac{1}{\epsilon^2}\right)^2$. Then by using again Lemma 2.7 and (10)

$$\begin{aligned} M_\epsilon[r_\epsilon] \geq M_\epsilon[s_\epsilon] &= (2k + o(1))\alpha_{\epsilon, C_k}(\rho_\epsilon) + O(\epsilon^{2\tau}) \\ &= \epsilon^2(2kC_k + o(1)) \frac{1}{\left(\log \frac{1}{\epsilon^2}\right)^{N-2}}. \end{aligned} \tag{11}$$

We are going to prove that $\frac{r_\epsilon \beta_k}{\epsilon \log \frac{1}{\epsilon^2}} \to 1$ as $\epsilon \to 0$. By contradiction assume that there exists a sequence $\epsilon_n \to 0$ such that $\frac{r_{\epsilon_n} \beta_k}{\epsilon_n \log \frac{1}{\epsilon_n^2}} > 1 + c$. Using once more (10)

$$\begin{aligned} M_{\epsilon_n}[r_{\epsilon_n}] &\leq \epsilon_n^2 (2kC_k + o(1)) \frac{1}{\left| \frac{r_{\epsilon_n}}{\epsilon_n} \beta_k \right|^{N-2}} + O(\epsilon_n^{2\tau}) \\ &\leq \epsilon_n^2 (2kC_k + o(1)) \frac{1}{(1+c)^{N-2}} \frac{1}{\left| \log \frac{1}{\epsilon_n^2} \right|^{N-2}} \end{aligned}$$

which contradicts the (11). Now assume the existence of a sequence $\epsilon_n \to 0$ such that $\frac{r_\epsilon \beta_k}{\epsilon \log \frac{1}{\epsilon^2}} < 1 - c$. Then by the decay of w (observe that the function $x \mapsto \frac{e^x}{x^{\frac{N-3}{2}}}$

is nondecreasing for x large)

$$\begin{aligned}
M_{\epsilon_n}[r_{\epsilon_n}] &= (2k+o(1))\left(-\gamma_0 w\left(\frac{r_{\epsilon_n}}{\epsilon_n}\beta_k\right)+C_k\epsilon_n^2\frac{1}{\left|\frac{r_{\epsilon_n}}{\epsilon_n}\beta_k\right|^{N-2}}\right)+O(\epsilon_n^{2\tau})\\
&= (2k+o(1))\left(-\gamma_0 A_N\frac{e^{-\left(\frac{r_{\epsilon_n}}{\epsilon_n}\beta_k\right)}}{\left|\frac{r_{\epsilon_n}}{\epsilon_n}\beta_k\right|^{\frac{N-1}{2}}}+C_k\epsilon_n^2\frac{1}{\left|\frac{r_{\epsilon_n}}{\epsilon_n}\beta_k\right|^{N-2}}\right)+O(\epsilon_n^{2\tau})\\
&\le (2k+o(1))\left(-\gamma_0 A_N\frac{e^{-\left(\frac{r_{\epsilon_n}}{\epsilon_n}\beta_k\right)}}{\left|\frac{r_{\epsilon_n}}{\epsilon_n}\beta_k\right|^{N-2}}+C_k\epsilon_n^2\frac{1}{\left|\frac{r_{\epsilon_n}}{\epsilon_n}\beta_k\right|^{N-2}}\right)+O(\epsilon_n^{2\tau})\\
&= \epsilon_n^2(2kC_k+o(1))\left(-\frac{\gamma_0 A_N}{\epsilon_n^2 C_k}\frac{e^{-\left(\frac{r_{\epsilon_n}}{\epsilon_n}\beta_k\right)}}{\left|\frac{r_{\epsilon_n}}{\epsilon_n}\beta_k\right|^{N-2}}+\frac{1}{\left|\frac{r_{\epsilon_n}}{\epsilon_n}\beta_k\right|^{N-2}}\right)+O(\epsilon_n^{2\tau})\\
&\le \epsilon_n^2(2kC_k+o(1))\frac{1}{\left|\log\frac{1}{\epsilon_n^2}\right|^{N-2}}\left(-\frac{\gamma_0 A_N}{C_k}\frac{\epsilon_n^{-2c}}{|(1-c)|^{N-2}}+\frac{1}{|\beta_k^2|^{N-2}}\right)
\end{aligned}$$

which is in contradiction with (11) since $\left(-\frac{\gamma_0 A_N}{C_k}\frac{\epsilon_n^{-2c}}{|(1-c)|^{N-2}}+\frac{1}{|\beta_k^2|^{N-2}}\right)<1$ for n big. □

References

[1] A. Ambrosetti, *On Schrödinger–Poisson systems*, Milan J. Math. 76, (2008), 257–274.

[2] A. Ambrosetti and A. Malchiodi, *Perturbation Methods and Semilinear Elliptic Problems on $\mathbb{R}^N$*, Birkhäuser Verlag, 2005.

[3] A. Ambrosetti and D. Ruiz, *Multiple bound states for the Schrödinger–Poisson problem*, Comm. Contemp. Math. 10 (2008), 391–404.

[4] A. Azzollini and A. Pomponio, *Ground state solutions for the nonlinear Schrödinger–Maxwell equations*, J. Math. Anal. Appl. 345 (2008), 90–108.

[5] V. Benci and D. Fortunato, *An eigenvalue problem for the Schrödinger–Maxwell equations*, Top. Meth. Nonlinear Anal. 11(2) (1998), 283–293.

[6] V. Benci and D. Fortunato, *Solitary waves of the nonlinear Klein–Gordon equation coupled with the Maxwell equations*, Rev. Math. Phys. 14(4) (2002), 409–420.

[7] R. Benguria, H. Brezis and E.H. Lieb, *The Thomas–Fermi–von Weizsäcker theory of atoms and molecules*, Comm. Math. Phys. 79 (1981) 167–180.

[8] I. Catto and P.L. Lions, *Binding of atoms and stability of molecules in Hartree and Thomas–Fermi type theories. Part* 1: *A necessary and sufficient condition for*

the stability of general molecular system, Comm. Partial Differential Equations 17 (1992) 1051–1110.

[9] T. D'Aprile and D. Mugnai, *Non-Existence results for the coupled Klein–Gordon–Maxwell equations*, Adv. Nonlinear Stud. 4 (2004), 307–322.

[10] T. D'Aprile and D. Mugnai, *Solitary waves for nonlinear Klein–Gordon–Maxwell and Schrödinger–Maxwell equations*, Proc. Roy. Soc. Edinburgh Sect. 134A (2004), 893–906.

[11] T. D'Aprile and A. Pistoia, *On the number of sign-changing solutions of a semiclassical nonlinear Schrödinger equation*, Adv. Differential Equations 12 (2007), no. 7, 737–758.

[12] T. D'Aprile and J. Wei, *Standing waves in the Maxwell-Schrödinger equation and an optimal configuration problem*, Calc. Var. 25 (2005), 105–137.

[13] T. D'Aprile and J. Wei, *On bound states concentrating on spheres for the Maxwell–Schrödinger equation*, SIAM J. Math. Anal. 37 (2005), 321–342.

[14] B. Gidas, W.M. Ni and L. Nirenberg, *Symmetry and related properties via the maximum principle*, Comm. Math. Phys. 68 (1979), no. 3, 209–243.

[15] I. Ianni, G. Vaira, *Non-radial sign-changing solutions for the Schrödinger–Poisson problem in the semiclassical limit*, preprint.

[16] H. Kikuchi, *On the existence of solutions for elliptic system related to the Maxwell–Schrödinger equations*, Nonlinear. Anal. 67 (2007), 1445–1456.

[17] E.H. Lieb, *Thomas–Fermi and related theories and molecules*, Rev. Modern Phys. 53 (1981) 603–641.

[18] P.L. Lions, *Solutions of Hartree–Fock equations for Coulomb systems*, Comm. Math. Phys. 109 (1984) 33–97.

[19] P. Markowich, C. Ringhofer and C. Schmeiser, *Semiconductor Equations*, Springer-Verlag, New York, 1990.

[20] D. Ruiz, *Semiclassical states for coupled Schrödinger–Maxwell equations: concentration around a sphere*, M3AS 15 (2005), 141–164.

[21] D. Ruiz, *The Schrödinger–Poisson equation under the effect of a nonlinear local term*, J. Funct. Anal. 237 (2006), 655–674.

[22] J.C. Slater, *A simplifcation of the Hartree-Fock method*, Phys. Review 81 (1951), 385–390.

Giusi Vaira
Dipartimento di Scienze di Base e Applicate
per l'Ingegneria, sezione di Matematica
Università La Sapienza di Roma
Via A. Scarpa 14
I-00185 Roma, Italy
e-mail: `giusi.vaira@sbai.uniroma1.it`

MIX
Papier aus verantwortungsvollen Quellen
Paper from responsible sources
FSC® C105338

If you have any concerns about our products,
you can contact us on
ProductSafety@springernature.com

In case Publisher is established outside the EU,
the EU authorized representative is:
Springer Nature Customer Service Center GmbH
Europaplatz 3, 69115 Heidelberg, Germany

Printed by Libri Plureos GmbH
in Hamburg, Germany